THE FUTURE OF SATELLITE COMMUNICATIONS

THE FUTURE OF SATELLITE COMMUNICATIONS

George A. Codding, Jr.

Westview Press
BOULDER, SAN FRANCISCO, & OXFORD

Westview Special Studies in Science, Technology, and Public Policy

This softcover edition is printed on acid-free paper and bound in library-quality, coated covers that carry the highest rating of the National Association of Textbook Administrators, in consultation with the Association of American Publishers and the Book Manufacturers' Institute.

Copyright © 1990 by Westview Press, Inc.

Published in 1990 in the United States of America by Westview Press, Inc., 5500 Central Avenue, Boulder, Colorado 80301, and in the United Kingdom by Westview Press, Inc., 36 Lonsdale Road, Summertown, Oxford OX2 7EW

Library of Congress Cataloging-in-Publication Data
Codding, George A. (George Arthur)
   The Future of satellite communications / George A. Codding, Jr.
      p. cm.—(Westview special studies in science, technology,
and public policy)
   Bibliography: p.
   Includes index.
   ISBN 0-8133-8003-0
   1. Artificial satellites in telecommunication.
2. Telecommunication systems. I. Title. II. Series.
TK5104.C58   1990
384.5′1—dc20   90-30618
                                                                          CIP

Printed and bound in the United States of America

The paper used in this publication meets the requirements
of the American National Standard for Permanence of Paper
for Printed Library Materials Z39.48-1984.

10   9   8   7   6   5   4   3   2

CONTENTS

FOREWORD

Professor Codding was one of the principal intellectual leaders of a telecommunications policy research project that got under way at the University of Colorado in 1987, supported by a grant from the National Aeronautics and Space Administration.* He recommended that we begin our work with review of the present situation and emerging trends in telecommunications so that our policy studies would be well grounded. That was the genesis of this book.

Originally we were mainly interested in satellite communications and the interaction with Integrated Services Digital Networks (ISDNs), which were under development. The coverage has been widened and updated in this volume. In addition this work looks more to the future.

The result is a book that should be broadly useful to all interested in satellite communications. As a spin-off of work originally done for NASA, it represents the kind of fruitful synergism that can result from the application of university talents to practical policy problems.

We are grateful that NASA made the effort possible by funding the original work at the Center for Space and Geosciences Policy.

Radford Byerly, Director
Center for Space and
 Geosciences Policy
University of Colorado
Boulder, Colorado

---

* Radford Byerly, Frank Barnes, George Codding, Jefferson Hofgard, *NASA and the Challenge of ISDN: The Role of Satellites in an ISDN World*, NASA Grant NAGW-1105 (Boulder, CO: May 25, 1989).

ACKNOWLEDGMENTS

Special thanks go to three individuals in the University of Colorado Center for Space and Geosciences Policy: Rad Byerly, Director, who provided unflagging enthusiasm and support; Jeff Hofgard, Rad's hard-working assistant; and Darlene Jeune, the Center's constantly cheerful and extremely competent secretary. Professors Frank Barnes and Warren Flock of the CU-Boulder Interdisciplinary Telecommunications Program made special contributions, as did a number of individuals who were students in that Program. Charles Sheridan, one of those students, worked for me for more than a year on research and editing. A great deal of assistance was provided by students David Fahey, Paul Ferris, Marty McClosky, Maria Pardee, and Michael Seebeck. Mention should also be made of the help of Art Rholoff, Richard Andrews, and Josephine Squires, Department of Political Science graduate students. In addition, a note of thanks is due to Joseph Pelton, new Director of the Interdisciplinary Telecommunications Program, who read the manuscript and made a number of excellent suggestions.

George A. Codding, Jr.

INTRODUCTION

Little more than three decades after the first successful launching of an artificial satellite, communication satellites have become the foremost instrumentality of long-distance international communication on a global scale. In 1988, for instance, INTELSAT could claim that its thirteen operational satellites were linking "172 countries, territories and dependencies around the globe via 1,738 full-time earth station-to-earth station pathways (exclusive of transponder leases) among 767 earth station antennas for a variety of voice, video, data, and audio services."[1] INTERSPUTNIK was performing a similar service for its fourteen members and nine associated countries of the Socialist and Communist Bloc, and EUTELSAT, ARABSAT, and to some extent the Indonesian Palapa system were fulfilling the telecommunications needs of their respective regions.[2]

Since the mid-1970s satellites have also become an important element in domestic communications. The United States, USSR, and Canada all have extensive systems of domestic satellite communications, as do Australia, Brazil, France, India, Indonesia, Mexico, and Japan. Other countries, such as the Federal Republic of Germany, Italy, and China have domestic systems in the planning stage. In addition, INTELSAT provides domestic satellite services to another twenty-eight countries, including Algeria, Argentina, Chile, China, Colombia, Côte d'Ivoire, Denmark, France, Gabon, Germany (FR), India, Italy, Libya, Malaysia, Morocco, Mozambique, New Zealand, Nigeria, Norway, Peru, Saudi Arabia, South Africa, Spain, Sudan, Taiwan, Thailand, Venezuela, and Zaire.[3]

Many new uses for satellite communication, in addition to broadcasting and fixed services, are also becoming available. INMARSAT, for instance, while steadily extending its maritime mobile service, is also inaugurating an aeronautical mobile service and experimenting with a land mobile service. Direct broadcasting by satellite of sound and television directly into the home is slowly becoming a reality, although largely outside the U.S.

Recently, however, advances in telecommunication technology have introduced a major competitor to satellites, the fiber optic cable. Just as special characteristics, including broadcast capability and distance insensitivity, allowed satellites to compete successfully with earlier methods of long-distance communication, such as standard radio and the older underseas cable, fiber optic cable threatens to become a strong competitor on many of the long-distance communication routes currently being served by satellites. Two optical fiber cables have been installed in the Atlantic to serve the lucrative U.S.-to-Europe route. One fiber optic cable now connects the U.S. with various Asian nations across the Pacific and another is under active consideration. Additional cables are being planned for the Atlantic, the Pacific, and the Caribbean. A number of countries are also turning to fiber optic cable for their long-distance domestic communication needs.

As a result, the exact role that satellites will play in the provision of long distance communication in the future is under question. The nature and extent of satellite applications will depend on developments in communications satellite technology as well as developments in fiber optic cable technology. The role will also be determined in part by the evolution of the cost structure involved in the utilization of the two methods of communication in response to technological advances.

The future role of satellite communications will be impacted by variables in addition to technology or economics, however, not the least important of which involves considerations of public policy at both the national and international levels.

In the near future the world's policy makers will have to answer a number of questions. Is it to the benefit of the individual nations and the world at large for fiber optic cable technology to replace satellite technology, or should regulations, standards, and policies be adopted that will guarantee that both technologies have a place in the overall scheme of things? Are there adequate space resources, especially orbital positions for geostationary communications satellites and the appropriate radio frequencies, for all who wish to use them? How should those resources be allocated? Are there any applications of satellite communication that should be outlawed, or at least regulated to a significant degree?

Although public policy is an essential element in the understanding of the future of satellite communication, it has been less extensively investigated than either technical or economic considerations. The purpose of this book is to explore these and related questions which the author feels will have an important impact on the future of satellite communication in our times.

The emphasis of the book will be on U.S. policy considerations inasmuch as the U.S. has been at the forefront of satellite technology and its application. The future of satellite communication will be dependent to a great degree on the role in information transfer that the United States gives to satellite communications.

After a brief review of the world of long-distance satellite communications and its fiber optic competitor, Chapter Two will address the impact of the earlier U.S. policy of global monopoly on the development of international satellite systems. Chapter Three will then explore the succeeding U.S. policy of global competition and various reactions to that policy. Chapter Four will discuss the policies and mechanisms involved in the allocation of scarce space resources, especially geostationary orbital positions and the appropriate radio frequencies, under the auspices of the International Telecommunication Union (ITU). Chapter Five will be devoted to a discussion of how the U.S. obtains its share of these scarce space resources and allocates them to government and private sector users. A final chapter will contain the conclusions and some speculation as to the future of international satellite communications networks.

NOTES

1. See INTELSAT, *INTELSAT Report, 1987-88*, Washington, D.C., 1988, p. 2.

2. Joseph N. Pelton and John Howkins, *Satellites International*, New York: Stockton Press, 1988, pp. 119-137 and 192-193.

3. *Ibid*, p. 23.

CHAPTER ONE

COMMUNICATIONS SATELLITES, SYSTEMS AND SERVICES

INTRODUCTION

Satellite systems and applications have developed in a relatively short time under the influence of such factors as technological innovation, cost advantages, and the needs of users. A knowledge of some of the more important aspects of satellite systems and their applications is thus an essential background to an understanding of the impact of domestic and international policy-making on future uses of satellite communications.

This chapter will provide an overview of current satellite communications services, a look at the systems that provide those services, and a discussion of some of the recent developments in satellite technology and its applications. Although the emphasis of this work is on satellites, it would obviously be incomplete without addressing the technology that has become its greatest rival. Accordingly, a section discusses the strengths and weaknesses of fiber optic cable. The chapter concludes with an estimation of the role that satellite communications is likely to play in the future.

COMMUNICATIONS SATELLITE SERVICES AND SYSTEM ARCHITECTURES

Types of Services

The International Telecommunication Union (ITU) recognizes twelve categories of communications satellite services:[1]

1. fixed satellite service
2. broadcasting satellite service
3. mobile satellite service
4. radio determination satellite service
5. space operation service
6. space research service
7. earth exploration satellite service
8. meteorological satellite service
9. inter-satellite service
10. amateur satellite service
11. radio astronomy service
12. standard frequency and time signal satellite service

This book focuses on the issues surrounding satellites which provide fixed, broadcast, mobile, and radio-determination services, the most prominent services in use today. It should be noted that since satellites are equipped with multiple transponders and often have dual frequency capabilities, two or more of these services may be provided by the same satellite. In particular, many satellites currently in orbit provide both fixed and broadcast services and, depending on the definition, some also offer direct broadcast services.

The *fixed satellite service* (FSS) relays voice or data, or distributes radio or television signals, from one point to another (point-to-point) or to many ground stations (multipoint). A *broadcast satellite service* (BSS), or *direct broadcast satellite* (DBS) in popular terminology, beams radio and TV program signals from originating stations directly to small, low-cost, home-mounted terminals via high-powered satellite transmitters.[2]

Mobile satellite services (MSS) permit communications with and between earth stations which are in motion. Mobile satellite systems are used in shipping, aviation, interstate trucking, railroading, medical emergencies, natural disasters, and planned temporary installations (such as at sites for mining or oil and gas drilling). They are also used to provide communication in rural and other lightly populated areas. These areas of application are typically remote, sparsely populated, or characterized by communication facilities which are either inadequate, unreliable, or non-existent.

Both the United States and Canada are heavily involved in mobile satellite technology. The U.S. program, called the Mobile Satellite Experiment (MSAT-X), was initiated in 1984 and is managed for the National Aeronautics and Space Administration by the Jet Propulsion Laboratory in Pasadena, California. The purpose of the MSAT-X program is to support U.S. industry in its efforts to initiate commercial mobile satellite services. To this end, the MSAT-X program concentrates on developing advanced ground segment technologies and techniques for mobile communications in future generation high-capacity systems. Areas of concentration include vehicle antennas, mobile radios, network management, and multiple-access systems.

The Canadian government's Department of Communications is actively promoting the development of a mobile satellite system for Canada (MSAT) in parallel with the U.S. NASA-based MSAT-X project. Canada's MSAT has the potential to become the second most important communication network in Canada after the fixed telephone network. The MSAT system is designed to permit any combination of mobile and fixed connections with the only limiting factors being the cost of user terminals and the intended use of the assigned frequency spectrum. In this context the following major service categories have been defined as:[3]

1. Satellite mobile radio service to provide private communications between mobile units or between units and a base station.
2. Satellite mobile telephone service to provide two-way radio telephone communications between mobile units and the public telephone network or between mobile units.
3. Remote telephone service to extend telephone service to some of the more than 100,000 Canadian households who currently live beyond the reach of the existing public telephone network, bringing "thread of life" communications to these homes and to temporary or seasonal dwellings like a field camp or family cottage.
4. Satellite mobile data services to give subscribers with mobile data units dial-up access to computer data bases. Subscribers dial directly into computer banks without an intermediary, and the video screen of their mobile data units displays information which may be entered or processed. Data can be encoded to ensure confidentiality.
5. Data acquisition and control to allow for data collection from remote monitoring and alarm devices and transmission of commands to automated control stations. Possible industrial applications include the monitoring and control of pipelines, railways, power lines, and oil wells.
6. Paging services nationwide to send one-way messages to any vehicle, ship, or aircraft located anywhere in Canada.

The United States now has two mobile satellite systems in place and one in an advanced planning stage. Geostar of Washington, D.C., and Qualcomm of San Diego are both offering one-way communication with trucks and both are planning to offer two-way communication in the near future. The Geostar system has been installed on more than 1,000 vehicles and Qualcomm has contracts for more than 7,500 terminals. Both systems offer their service via leased channels on GTE Spacenet satellites. On May 31, 1989, the FCC adopted an order granting a license authorizing American Mobile Satellite Corporation (AMSC) to offer, on a common-carrier basis, a full range of land, aeronautical, and maritime mobile satellite services.[4] AMSC plans to operate three satellites at 62° W, 101° W, and 139° W, using the 1 545-1 559 MHz and 1 646.6-1 660.5 MHz bands.[5]

Radiodetermination Satellite Service, or RDSS, is essentially a set of radiocommunication and computational techniques that enables users to determine their geographical position precisely and to relay this and similar digital information to any other user. The practical applications of RDSS are many. A land-based RDSS can be used to keep track of a fleet of trucks, automobiles, or railway cars; ascertain their geographic location at any time; and send messages to the vehicles while they are en route. In aeronautical applications, RDSS can provide navigational information. The pilot of an aircraft can access the system to obtain information on the location of obstacles such as radio towers or mountain ridges. RDSS can also provide information on the location of landing strips. The maritime applications of an RDSS include fleet tracking, surveying, search and rescue planning control, boat dispatching, and fishing area identification.[6]

Applications, Domestic and Foreign

The FCC estimates that telecommunications accounts for 5%, or $200 billion, of the U.S. GNP. This amount is split equally between broadcast and common carrier services. The satellite industry accounts for $500 million annually of telecommunications expenditures, based on the annual cost of deploying and maintaining domestic satellite transponders and the 6,000 licensed and one million unlicensed ground terminals. If one were to take into account the economic function that satellite services play in the cable television and broadcasting industry, and the number of new services that satellite communications create, it would certainly be more than $500 million annually.[7]

Users of domestic telecommunications services in the United States enjoy access to a variety of competing carriers, employing both satellite and terrestrial transmission systems, to meet their needs for voice, video, and data transmission services.

The United States is currently served by thirty-three communications satellites.[8] The country already had a highly-evolved terrestrial public switched telephone network (PSTN) before the arrival of satellite communications. As a result, ordinary voice traffic on satellites was relegated to serving only those users not easily tied into the PSTN, such as TV distribution in "thin route" locations or areas where the terrain makes land-based connections very expensive. Not surprisingly, then, voice traffic on U.S. domestic satellites is third in traffic volume, behind television distribution and data communications. Cable TV programmers pioneered the use of satellites for program distribution, and such traffic constitutes about half of the television service on satellites. The usage by broadcasters and closed circuit systems is more recent, but it is growing rapidly.

Referring to Table 1.1, we find that over 40% of the U.S. transponders in use are devoted to television transmission, about 31% handle data traffic, or very small aperture terminal (VSAT) networks for business, and about 25% are for voice traffic. It should be noted that voice traffic has decreased since 1986.

Much of the video and data traffic is carried on private networks whose owners sell satellite transponder capacity to businesses. Satellite networks in the U.S. may also be accessed directly, provided that the user has a satellite station on-site. Users can access the networks of satellite service vendors when their services are required, or users may own or lease individual transponders. In the U.S., the number of private satellite networks devoted to internal corporate communications jumped from eighty-nine in 1987 to 143 in 1988. TV and VSAT applications are cited as the chief cause of this expansion.[9] Several businesses listed 2,000 or more satellite sites. The Farmers Insurance Group of Los Angeles reported a satellite data network of 2,500 sites.[10]

The U.S. traffic mix contrasts sharply with that of the INTELSAT system, the world's largest provider of domestic and international satellite services. As seen in Table 1.2, voice traffic accounts for almost 70% of INTELSAT's full-time international traffic.

Links, Frequencies, and Power

A communications satellite functions as a radio link between one or more variously dispersed earth stations. The radio link from an earth station transmitting to a satellite is referred to as an uplink,

Table 1.1

**Estimated U.S. Transponder Usage by
Traffic Type, March, 1986**

Traffic Type	C band	K band	Totals
Television			
Cable TV	68	1	69
Broadcast	43	17	60
Closed Circuit	9	3	12
Total TV			**141**
Radio	**5**	**0**	**5**
Voice			
Heavy Route	7	0	7
Light Route	37	0	37
Private Networks	39	1	40
Total Voice			**84**
Data (Private Networks)	72	29	101
Active Totals	**280**	**51**	**331**
Inactive Totals	**133**	**55**	**188**
TOTALS	**413**	**106**	**519**

Source: Andrew Inglis, "The United States Satellite Industry -- An Overview," *Conference Proceedings, Fibersat 86* (Vancouver, Canada: Wescom Communications Studies and Research Ltd., 1986), p. 74. The data in this table are adapted from the semi-annual FCC Transponder Loading Report which does not record usage type directly. Consequently, these figures should be considered as estimates only.

Table 1.2

INTELSAT Revenue by Traffic Type in 1988

Type	Percent
Full-Time Analog Service..	69.0
Digital Services..	13.3
Occasional Use Television...	4.8
Domestic Telecommunication Services.............................	4.1
International Television Leases.......................................	3.3
Other..	5.5
Total...	100.0

Source: INTELSAT, *INTELSAT Report, 1987-88*, Washington, D.C., 1988, pp. 38-39.

Table 1.3

Satellite Communication Frequency Bands*

L band, 1-2 GHz
S band, 2-4 GHz
C band, 4-8 GHz
X band, 8-12.5 GHz
Ku band, 12.5-18 GHz
K band, 18-26.5 GHz
Ka band, 26.5-40 GHz

* Above 40.5 GHz most of the assignments are used only for experimental purposes.

Source: Adapted from *Intermedia*, July/September 1989, p. 51.

whereas a downlink is the link going from a satellite to a receiving earth station. In effect, a satellite link is "a distant microwave repeater which receives uplink transmissions and provides filtering, amplification, processing, and frequency translation to the downlink band for retransmission."[11]

The offsetting of uplink and downlink frequencies, known as frequency translation, is necessary to prevent mixing and interference between uplink and downlink frequencies and signals within a single satellite transponder. The most commonly used frequency bands are the C, Ku, and Ka bands. Satellite communications in the C band employ uplink frequencies of 6 GHz and downlink frequencies of 4 GHz. Likewise, the Ku band matches 14 GHz uplinks with 12 GHz downlinks, and the Ka band pairs frequencies of 30 and 20 GHz respectively.[12]

Each band has certain advantages. Links in the C band provide low loss due to rain (for earth station angles above 5°). The Ku band requires a significant fading margin to contain severe signal attenuation due to rain, but increasing problems of frequency interference in the congested C band have made its use, and that of the Ka band, more and more necessary. The higher frequency bands are ideal for spot beams and tend to require smaller and more compact earth station antennas. In both the Ka and Ku bands, paths at elevation angles below about 5° experience more fading than paths at high elevation angles.

Transmission in a satellite system is described in terms of channels, circuits, and half-circuits. A channel allows transmission in one direction only, from a ground station, across a satellite, and to one or more receiving ground stations. A circuit is a channel with two-way communications ability. A half-circuit is a two-way link between a satellite and a single ground station. A half-circuit is more of an accounting unit than a technical term.

The measure of satellite capacity depends on the particular application. Circuits are the practical unit of measure for common carriers providing voice communication which is, of necessity, two-way. On the other hand, data and television transmission is typically one-way and thus measured in channels.[13]

The strength of a transponder's signal decreases as the distance between the receiver and the center of the beam increases. The effective isotropic radiated power of a satellite transponder, or EIRP, provides the basis for calculating the intensity of a satellite signal. Decibels referenced to one watt of power, dBWs, are the usual unit measure of EIRP. The EIRP at the center of the beam of a transponder is found by adding the gain of its transmitting antenna in dB to the power level of the transponder's traveling wave tube amplifier in dBW and by then subtracting losses occurring between the antenna and the amplifier in dB.[14]

Orbits

Satellites which provide telephony services almost always reside in a geostationary satellite orbit (GSO), 22,238 miles (35,786 km) above the earth. At this altitude, a satellite can follow the earth's rotation while remaining stationary with respect to the earth's surface. Meteorological and remote sensing satellites also benefit from GSO's fixed view of the earth.

For satellite applications such as reconnaissance, weightlessness or vacuum experiments, material processing and space manufacturing activities, and localized weather and resource imaging, the low earth orbit, less than 500 miles high, is the easiest and most economical alternative to the geostationary orbit. The U.S. space shuttle can launch satellites directly into low earth orbit, as well as retrieve them from there. Additional capability, however, is needed to reach medium altitude orbit or GSO.

Satellite applications which require a greater view of the earth, such as navigation satellites, follow the medium earth orbit, 7,000 to 8,000 miles (11,200 to 12,800 km) above the earth. When launched into a polar orbit a satellite can monitor the earth's surface continuously and circle the earth in 90 minutes. The medium earth orbit is especially useful for navigation, as well as remote sensing and weather monitoring.

The super-synchronous orbit is a catch-all term for orbits above the geostationary orbit and below that of the moon. It is difficult to locate satellites in this orbit, thus it is of interest mainly to the military. Scientific applications could also benefit from the super-synchronous orbit's proximity to the Lagragian points, where the gravitational effects of the moon and earth are in balance.[15]

ADVANCED TECHNOLOGIES AND
THEIR APPLICATIONS

This section provides an introduction to some of the recent advances in satellite communications technology and applications. Descriptions of the INTELSAT VI and ACTS satellites and VSATs will be used to exemplify the trends in technology. An outline of techniques used to improve the switching capacity and transmission quality of satellite communications follows. Trends in satellite applications conclude this section.

Certain portions of the material that follows are a bit technical.

If the reader should find it heavy going, he or she is advised to proceed directly to The Major Advantages of Satellite Communication, which starts on page 24.

Hardware and Switching Techniques

Advanced Communications Technology Satellite (ACTS) Program

The ACTS program was conceived in 1979 and is sponsored by NASA's Office of Space Sciences and Applications and managed by NASA's Lewis Research Center. Its dual goals are to support continued U.S. industry leadership in the world communications satellite market and to encourage the use of the technologies which it is developing. The satellite will operate digitally and its sophisticated technical features include electronically hopping spot beam antennas, frequency reuse, on-board processing and switching, demand assignment multiple access (DAMA), Ka band transmission, and small ground terminals. The research and development objectives of the ACTS program are directed toward developing technology which is too technically demanding, costly, and risky to be left to the private sector alone. ACTS is scheduled for a May, 1992, launch date, and NASA anticipates an operational life of from two to four years. The agency intends to demonstrate the feasibility of ACTS technologies and applications through evaluation and applications experiments, encouraging as much private, government, DOD, and university involvement as possible.[16] The most important technical features of ACTS are presented below.

Electronically Hopping Spot-Beam Antennas. The footprint of domestic satellites typically covers all or a large part of the continental U.S. Although such a wide footprint optimizes the coverage of a satellite, it reduces the effective isotropic radiated power (EIRP) of satellite signals. The ACTS program intends to boost EIRP by concentrating its beams on specific regions. NASA expects EIRP improvements of up to 20 dB over the 27 to 33 dB antenna gain produced with beams which encompass the continental U.S.

Three spot beams will remain fixed on Cleveland, Atlanta, and Tampa, while two other spot beams will hop between a total of 13 discrete locations and any locations within two regions known as the east and west sectors. ACTS will cover 20% of the continental U.S., a scope sufficient for the program's experimental mission. An additional mechanically steerable antenna will employ a 1° beam capable of providing uplink and downlink capability anywhere on the disk of the earth within the view of the ACTS's 100° west position.

The hopping beams will be capable of focusing on their target areas for variable durations, thereby optimizing the use of satellite capacity by adapting coverage to fluctuations in regional traffic demand. For regions of stable and high traffic demand, stationary beams will enable a simpler approach to beam distribution.[17]

The total channel capacity of a satellite can be increased by a technique known as frequency reuse. ACTS beams will feature frequency reuse by means of spatial and polarization diversity techniques. Spatial frequency reuse coordinates multiple directional beams which operate at the same frequency. Polarization frequency reuse implements orthogonal polarization at the same frequency on multiple beams.[18] Polarization diversity in ACTS will be applied not only between beams, but also between the uplink and downlink of polarized beams.[19]

Interbeam Communications Through On-board Processing and Switching. Two on-board beam-to-beam routing mechanisms will add interbeam communications to ACTS capabilities. The satellite provides digital transmission and uses time division multiple access (TDMA), a transmission technique in which earth stations take turns sending data through a common transponder. The use of TDMA implies a digital communications system capable of switching digitized voice, data, or video in an integrated fashion.[20] ACTS will incorporate two modes of TDMA, a baseband processor mode and a microwave switched matrix mode. Both modes will allow single-hop satellite switching and direct terminal-to-terminal connection, eliminating the intermediary ground switching hubs used in current satellite systems.[21]

By allocating transmission facilities when and where they are needed, the ACTS will give users the capability to dynamically reconfigure their communications networks. Satellites incorporating this function will have the ability to handle a wide range of traffic loads, thereby meeting requirements for both high-volume trunk routes and relatively thin-route links. Users who require network services with the flexible interconnection of a large number of sites, particularly on an international basis, will clearly find advantages in the flexibility that such communication satellite service offers.[22]

On-board stored baseband switched time division multiple access (OSBS/TDMA) regenerates and stores baseband signals on board the ACTS for the two hopping beams. The satellite's processor includes a baseband switch which routes stored data from input to output storage locations, thus facilitating interconnection of uplink and downlink beams connected to the storage locations. The modulation technique used to load the stored signals onto the beams is known as serial minimum shift keying, or SMSK.

The three stationary ACTS beams will achieve interconnection through a microwave switched matrix. The microwave switching, also known as IF, will use satellite switched time division multiple access techniques, or SS/TDMA. SS/TDMA requires no on-board storage or processing other than switching, and, since no on-board signal regeneration is necessary, ground terminals will be free of any satellite-imposed limits on modulation techniques. One or both of the hopping spot beams will also be able to access the SS/TDMA mode, as the system can handle any three of the five possible beams.[23] A switching system which incorporates SS/TDMA functions as a distributed time-space-time switch. Transmitting ground terminals operate as time switches by sequencing traffic bursts so that they arrive at the microwave switch matrix during the interval when it connects to the downbeam which broadcasts to the terminals which receive the data burst. The microwave switch matrix operates as a space switch, since it provides dynamic connection between uplink and downlink beams. The receiving terminals functions as the second time switch, opening its aperture and selecting a channel to receive the ACTS downlink beam.

The relative advantages of SS/TDMA and OSBS/TDMA modes depend on system traffic patterns. SS/TDMA is best suited to situations where interbeam traffic is heavy and limited to less than 100 ground terminals. If the traffic is light and involves 100 or more terminals, system overhead in the form of data communication preambles and guard times passed between transmitting and receiving terminals becomes inefficient. In such a case, OSBS/TDMA is the more advantageous choice, since its on-board memory performs all time and space switching requirements. Terminals are freed of having to coordinate their operation within the intervals of a microwave switch matrix and can broadcast data bursts at will. Placing all system switching in the satellite of course adds cost and complexity to the space segment, but in a light-traffic network with a more than 100-terminal situation, the savings due to network efficiency makes OSBS/TDMA a more economical choice than SS/TDMA.[24]

One of the key technical objectives of the ACTS program is to develop and demonstrate the utility of effective satellite communications in the Ka band (30/20 GHz). This will open up a largely vacant range in the frequency spectrum.[25]

Because the OSBS/TDMA performs all message-switching functions on board the satellite, it is the key technology permitting small terminals to exchange all-digital voice, data, and video messages in a single hop without any terrestrial connections. The OSBS/TDMA is unique in using onboard forward error correction and burst rate reduction to compensate for the rain fade prevalent at 30/20 GHz.

This compensation minimizes radio frequency power requirements, thereby reducing the cost and size of ground terminals.

Of particular interest are two earth terminals designed, built, and tested by Scientific Atlanta: a 2.4 meter diameter low burst-rate terminal and a 4.72 meter diameter high burst-rate, step-tracking terminal. Both terminals are installed and working at NASA's Lewis facility and will be used for ACTS experimental operations. The terminals are designed to transmit at 27.5 to 30.0 GHz and to receive at 17.7 to 20.2 GHz. Both models are designed to accommodate close satellite spacing and maintain critical pointing accuracy in high wind conditions.[26]

Scientific Atlanta also manufactures 1.8 and 2.8 meter diameter Ku band ground terminals that can be modified and adapted to satisfy ACTS experimental interests and requirements in the Ka band. These terminals are designed to perform in a wide range of commercial applications and to offer a low-cost approach for appropriate ACTS requirements and conditions.[27]

Another objective of the ACTS project is the development of the technology essential to intersatellite links for the next generation of communications satellites. The MIT Lincoln Laboratory was asked to design experiments for the U.S. Air Force Space Division. The experiments would involve a series of tests to verify the engineering parameters, operational performance, and technological maturity of laser communications. The need for higher data rates has been a key impetus for developing laser communications. Current intersatellite links are limited by antenna size to 50 to 100 Mbps (megabits per second). Intersatellite data rates of 2 to 4 Gbps (gigabits per second) may be needed in the future. Moving to higher frequencies, or using laser communications, offers a number of benefits: smaller antennas, wider bandwidth frequency slots, and fewer channel interference problems.[28]

Demand Assignment Multiple Access (DAMA)

This technique assigns satellite capacity according to call demand. By enabling users to access and release system capacity, DAMA optimizes the use of satellite capacity and is particularly advantageous to light traffic users. In a situation where four circuits would be necessary to attain a certain grade of service between two earth stations using conventional pre-assignment techniques, DAMA could reduce the circuit requirement to one, realizing a fourfold increase in satellite capacity.

In systems utilizing more than 100 terminals, DAMA techniques employ shared control channels known as request channels. Users send requests to an automatic network controller which then transmits channel assignment information to the user on a return channel. This is called a random access technique.

For systems of fewer than 100 terminals, the balance of system access control shifts a little closer to the terminals themselves. In the ACTS system, for example, the DAMA application will assign dedicated two-way control channels to all active terminals. For a terminal to achieve active status, it need only be logged on; it needn't be communicating or queued to communicate with other terminals. The control channels, or order wires, will provide continuous communications between active terminals and the ACTS control station. Control channels transmitting from terminals to the control center are known as inbound order wires, whereas the corresponding control channels going from the ACTS control center are dubbed outbound order wires. Order wires will also transmit other information, but more important is their role to maintain, receive, and transmit synchronization. In the ACTS system, another pair of control channels will connect the ACTS master control station and the on-board baseband processor (OBP).[29]

Although other countries have developed innovative experimental communications satellites, including Japan's CS-2 and BS-3, Italy's Italsat, and the European Space Agency's Olympus, ACTS is still at the forefront of world technology.

INTELSAT VI

The newest generation of satellites in the INTELSAT system, INTELSAT VI, had its first launch in late 1989. The five planned INTELSAT VI satellites will feature many technical enhancements and innovations.

With digital compression equipment, the fifty transponders of the INTELSAT VI will be capable of handling up to 120,000 simultaneous 2-way telephone circuits and at least three television channels. By comparison, the INTELSAT V-A satellite had an initial capacity of about 15,000 2-way telephone circuits and even with digital compression equipment will probably be able to carry little more than 50,000 voice circuits. The expected life of INTELSAT VI is thirteen years compared to the seven years of INTELSAT IV and V.[30]

In INTELSAT VI the C band frequencies are reused six times through two hemispherical beams and four zone beams using dual circular polarization. The Ku bands are reused twice by spatially isolated east and west pointing spot beams using linear orthogonal polarization.[31]

The INTELSAT VI satellites are concentrating their system cost and complexity in the space segment rather than on the ground. Early satellite communication networks did the reverse, employing satellites which transmitted relatively low-power signals whose reception required large, costly earth stations. All five INTELSAT VI satellites are being modified to increase their EIRP to 44 watts.[32] These high power output levels make possible direct access to the satellite by small, inexpensive earth stations. The cost of modifications to the INTELSAT VI series amounts to about $85 million.[33]

INTELSAT VI will also feature the first commercial application of satellite switched time division multiple access, allowing interconnection between the satellite's six receive antenna beams and six transmit antenna beams.

Very Small Aperture Terminals (VSATs)

Small satellite dishes, or VSATs, represent what could be the largest growth sector of the satellite industry in the coming years. Receive-only VSATs capable of handling data rates of 9.6 kbps can be as small as 75 cm in diameter and cost around $2,000, including a microprocessor and a printer. Receive-and-transmit VSATs allow direct access to satellites and can be purchased for about $5,000.

The low cost of the VSATs, achieved without reduction in reliability or performance, is due primarily to the extensive use of microprocessing. In receive-only VSATs, a single low-cost microprocessor performs five different major functions:[34]

1. automatic gain control, automatic frequency control, and automatic phase lock loop control;
2. forward error correction for correcting errors in data transmission;
3. packet demultiplexing for use in a packet switched network;
4. interface control for speed, code and protocol conversions handling both synchronous and asynchronous protocols and data rates in the range from 45 to 19.2 kbps;[35] and
5. diagnostic functions that monitor the performance of the earth station in detail and display the resulting status on a panel of light-emitting diodes. The built-in diagnostic capability also allows an end user to quickly install and orient the VSAT with a compass and an angle-finder.

Small and portable, VSATs are easily installed and can be used wherever there is a direct line of sight to a satellite. Connected to a data terminal through standard RS-232C or RS-422 interfaces, VSATs can be used to link a remote computer terminal to a communications network or provide access to a centralized computer or data base.

Currently, the largest market for VSATs resides in the establishment of private networks by large businesses, offering such services as: interactive data transmission, multipoint data transmission, 56 kbps switched facsimile, and point-of-sale terminal transactions. Direct end user connection to the satellite network achieves another economic benefit by permitting a user to bypass the local exchange carrier and avoid paying local loop or subscriber access charges. As mentioned earlier, VSATs are common hardware in the private satellite networks operating in the U.S. VSATs are ideal for networks involving a large number of geographically dispersed sites which have low volume or specialized traffic requirements.

Another potential market for VSATs is in receiving direct broadcast satellite (DBS) services for residential users. Predictions about the size of this market both inside and outside the U.S. vary greatly, and it remains to be seen if potential DBS service providers will be able to assemble the financing to "get off the ground" and compete with the cable TV industry for home delivery of entertainment and other services.

Teleports

As the name might imply, a teleport provides to the telecommunications industry what an airport provides to the airline industry. Teleports are a recent addition to the telecommunications industry; the first few became operational in the early 1980's. They are usually located near large urban areas. According to the World Teleport Association, a teleport is:[36]

an access facility to a satellite or other long-haul telecommunications medium, incorporating a distribution network serving the greater regional community and associated with . . . a comprehensive related real estate or other economic development.

Many other less formal but equally descriptive definitions abound, but each teleport is usually somewhat different, being influenced by the needs of the users in the region which it serves. Some teleports provide total end-to-end connectivity which can be leased on an hour-by-hour basis, catering to the needs of news services, for example. Others are equipped to provide flexible and cost-effective links to the international marketplace over extended periods of time. In the United States, a total of fifty-one teleports are in operation.[37] The World Teleport Association reports over seventy teleport projects either in operation or under way in twelve countries.[38] Whatever the location or specialty, teleports may assume an increasingly important role in providing a diverse range of satellite and other long-haul communications services to the U.S. and the international business community in the coming years.

Improvements in Satellite Switching and Transmission

Improvements in Signal Quality

Although the practical advantages of placing communications satellites at the geostationary altitude outweigh the disadvantages, such long transmission paths are the source of a few problems. A voice or data signal transmitted through a satellite in geostationary orbit has a travel time of approximately one quarter a second (250 milliseconds). In voice communications, this nearly quarter-second delay can be quite noticeable and distracting, although the results of several user surveys indicate that the adverse effects of delay do not significantly impair the quality of the communications.[39]

Transmission delay can also present problems in various data communication applications. Various techniques and network protocols have been developed which significantly reduce the problem of transmission delay in data communications.[40] However, it is more difficult to cope with delay in voice communications. Traveling one way only, video transmissions do not create any problems in this respect.

Transmission delay may lead to echo problems. Echo suppressor technology developed for early satellites was not very successful, but modern cancellation techniques effectively eliminate the problem of echoes.[41]

Error Detection and Correction

As with any other telecommunications transmission, the transmission of signals over a satellite communication system always results in some degradation of the quality of the information carried by the signal. Propagation disturbances, especially attenuation caused by rain on communication links operating at frequencies of 14/11 GHz and higher, contribute to the degradation of signal quality. In digital links the degradation of the information content of a signal is measured in terms of the bit error rate.

Various techniques for error detection and correction are used in digital communication systems. Error detection and correction techniques which rely on the retransmission of data are not practical on satellite links because of the long transmission delay. As a result, forward-error correction is generally preferred for satellite paths. In forward-error correction, extra "redundant" bits are added to the data stream. Their interpretation at the receiver enables an operator to know when and where an error occurs in received data. Errored bits can then be corrected at the receiving end of the transmission. Forward error correction techniques are capable of achieving extremely low bit error rates. The trade-off for this improvement in signal quality, though, is a reduction of the throughput of a data channel. In some forward error correction systems, the number of redundant bits is equal to the number of data bits, resulting in a halving of the data rate for a given channel transmission rate.[42]

Channel Capacity

Techniques for increasing the effective voice-carrying capacity of a communication system include adaptive pulse code modulation, speech interpolation and prediction schemes, and frequency reuse. These techniques can be employed in satellite links as well as in other transmission systems.

Adaptive pulse code modulation and delta modulation are techniques used to reduce the bandwidth required to transmit a voice signal. Substantial bandwidth savings are possible by using these techniques.[43]

The number of channels required for voice communications may be further reduced by the use of speech interpolation and prediction schemes that take advantage of the intermittent nature of voice communications. Originally developed for analog signals on submarine cable systems, the technique of reassigning channels during speech pauses is frequently called TASI, for time assigned speech interpolation. Digital implementations are called digital speech interpolation, or DSI. TASI takes advantage of the fact that during an average telephone call, a user listens about 40% of the time and talks about 40% of the time. The remaining 20% of the call is made up of pauses in the conversation. Each half of an occupied full-duplex telephone channel thus transmits speech for only about 40% of the time. When a talker pauses between words, or listens, a speech interpolation system takes away the channel and assigns it to someone else. When the talker again speaks, the system gives him or her a new channel.[44]

The INTELSAT VI satellites will use both adaptive differential pulse code modulation and DSI to increase their voice-carrying capacity to between three and five times that of a single voice channel.[45] The same ratio is true for the fiber optic transatlantic cables TAT 8 and TAT 9.

Satellites and ISDN

Any discussion of the role of satellites in the evolving global telecommunications network must inevitably address the concept of the proposed integrated services digital network (ISDN). The purpose of this section is to present the advantages of using communication satellites in the ISDN and consider some of the potential roles that satellites can play in realizing the full potential of an ISDN.

The ISDN has been presented as a global telecommunications network providing access and interconnectivity from any phone or data terminal on the planet to any other phone or terminal. Existing switching and transmission facilities and established data communication standards will be used whenever possible. It is reasonable to suggest that communications satellites can play a key role in the development of an ISDN by providing interconnectivity between many of the component parts of the network, much as they do today.

Founded on the concept of digital transmission of both voice and data, the ISDN is committed to providing end-to-end digital connectivity, a service that is presently available with digital communications satellites. Further, satellites using satellite switched/time division multiple access will have an enhanced ability to extend flexible networking and service features to ISDN users. Satellites can play a major role in the shaping of ISDN due to their unique capacity to offer both narrowband and wideband services to an almost infinite number of nodes in a non-hierarchical network and their ability to offer a truly global reach. But precisely how they will fit into the network is not entirely clear. A few possibilities deserve mention.

In the first option, a satellite network can be considered as connecting completely independent terrestrial networks. This may be necessary for two reasons. One, there is a growing apprehension that the ISDN may not emerge as a globally interconnected network but as a collection of fragmented, and perhaps nationally segmented, ISDNs. Digital communication satellites could function as interconnecting links between independent ISDNs in a network hierarchy where satellites are one level above the terrestrially-based ISDNs. Or, one may conceive of a situation where a public switched telephone network (PSTN) will continue to exist distinct from the ISDN but requiring interconnection with the ISDN. Again, satellites could be used to provide this link.

In the second option, satellite transmission would become a component of ISDN, but be restricted to the provision of transmission trunks on the network side of any network-provided switching facilities. In a sense, this does not involve compatibility at any of the seven open systems interconnection (OSI) model levels beyond the first because, in the provision of transmission trunks, a satellite network by itself has no interface with the user, and thus the question of compatibility with any of the OSI levels is not really relevant.[46] However, the satellite links involved would necessarily be in accord with the hierarchical digital rates and any performance-related ISDN recommendations. Similarly, the higher level protocols would of necessity be compatible with the delay present in satellite transmission.

The next level of penetration would be to provide "bearer services," i.e., providing the first three layers of OSI levels, the physical, data link, and network layers. In this case, the satellite network would interface with the user and might have to provide switched connections. Such a network could legitimately be termed an ISDN, and in this scenario multiple ISDNs could be envisioned.

Finally, one could imagine a satellite-based network which provides "teleservice" and is thus compatible with all seven OSI levels. There are today satellite-based networks which in the provision of enhanced services are akin to teleservices. The only reason they cannot be considered ISDN is that they do not follow the I-Series Recommendations of the International Telegraph and Telephone Consultative Committee (CCITT) of the International Telecommunication Union. This could change, as differences between such networks and ISDN standards are minor.[47]

THE MAJOR ADVANTAGES OF SATELLITE COMMUNICATIONS

The previous two sections have presented various aspects of satellite systems and services. This section summarizes the advantages that those aspects give satellite applications over other transmission media in the provision of traditional services and in meeting the requirements of offering new and innovative services.

To begin with, satellites are geographically robust. They provide economical solutions to the provision of basic telecommunication service in areas where terrestrial alternatives are either not available or not feasible due to geographical barriers, such as mountains or deserts. Satellites provide communications between isolated or remote areas and the rest of the world. The geographical robustness of satellite communications is most apparent in its ability to provide mobile services in any region within the footprint of a satellite. The use of VSATs in conjunction with satellites only enhances these geographical advantages.

Because satellites utilize dynamically assignable bandwidth capacity, they have an inherent ability to reconfigure communications networks and allocate network resources when and where they are needed. This gives users the flexibility to rearrange network configurations to adapt to variations in traffic load and directionality. It means, for example, that telecommunications managers can easily respond to the changing needs of either a growing business or a shifting population.

Satellites provide high-capacity links for use along high-volume routes such as trunk connections for international voice traffic. The dynamic network configuration allowed by satellites, combined with their high capacity, is particularly advantageous for point-to-multipoint wideband transmission for television broadcasting and tele-conferencing.

THE CHALLENGE OF FIBER OPTIC CABLE

Satellite technology has benefited from many impressive gains in the 25 years since INTELSAT I was launched in 1965. But other transmission media have enjoyed advances in technology as well. The most notable of these is fiber optic cable, which is finding wide acceptance in local network applications as well as in international voice and data transmission. Fiber is an attractive alternative to microwave and coaxial cable in terrestrial applications and to satellites in the transmission of transoceanic telephone traffic. In view of the completion of three major transoceanic fiber optic cables, and firm plans for more, it is worthwhile at this point to introduce the reader to fiber optic technology, its advantages over other transmission systems, and its expected growth and application.

Fiber Optic Technology and Its Advantages

Fiber optic technology provides telecommunications transmission by propagating light waves in the interior of glass or plastic fibers. Optical fibers do not conduct electricity. They combine three wrapped layers, a central core, a surrounding cladding, and an exterior shield.

Two kinds of optical fiber are used, multi-mode and single-mode.

Multi-mode fibers convey light in several modes of propagation and have cores with diameters several times the wavelength of the light they transmit. A representative multi-mode fiber would support light with a wavelength of 8 micrometers (or millionths of a meter), a core diameter of 50 micrometers, and a cladding thickness of 37.5 micrometers. The multi-mode fiber would have a total diameter of 125 micrometers. The light which it transmits refracts into several out-of-phase waves as it traverses the medium. Rays which enter a multi-mode fiber at different angles are refracted different numbers of times, and each entry angle is known as a mode of propagation, or "mode" for short.

Single-mode fibers have cores whose diameters are closer to the wavelength of the light which they transmit, and their carrying capacity is larger than the multi-mode optical fiber. They carry only a single ray, or mode, and are expected to dominate future fiber optic systems as their cost drops to that of multi-mode fibers.[48]

Like satellites, fiber optics enjoy a number of inherent technological advantages. They offer extremely high capacity in terms of bandwidth, are immune to electrical interference, and are very small and lightweight. They are not affected by radio frequency emissions from nearby communications systems, as are satellites and terrestrial microwave links operating at C and other bands.

Due to their low signal loss, significantly lower than that for copper wire and coaxial cable, the error rate for optical fibers is very low. Also in comparison with other enclosed transmission media, optical fiber can operate in high and low temperature environments without being adversely affected. Fiber optic systems also benefit from refinements in semiconductor technology, which enhances the capabilities of transmission and receiving equipment used in optical fiber networks.[49]

Fiber optics have two additional advantages over satellites: 1) they do not exhibit the transmission delays that are characteristic of geosynchronous satellites; and 2) their transmissions are generally considered to be more secure. Since fiber optic cables do not generate electrical fields, they must be tapped directly by someone who wants to intercept the signal. This is difficult although not impossible to do, and may be detected by sensitive monitoring devices.[50] By comparison, satellite transmissions are relatively easy to intercept, although it has been found that security can be maintained through the use of digital encryption. It is even possible to jam or force control of a satellite transponder.

The Advance of Fiber Optic Cable

Significant increases in the use of fiber optic cable have been occurring since 1983, when the first major long-haul fiber cable was installed between New York and Washington, D.C. 1986 saw U.S. interexchange carriers (long distance companies) install 34,000 miles of fiber optic cable, followed by 64,000 miles in 1987. It is estimated that 1.5 million miles of fiber optic cable will be installed during the 1990s.[51] It is expected that three-fourths of that cable will be owned and operated by AT&T, US Sprint, and MCI. The interexchange carriers are competing for access to major national and international routes, as well as for points of presence (a point where hook-up into a network occurs) close to Fortune 500 users.[52]

A breakdown of fiber optic cable investment among major U.S. interexchange carriers follows:[53]

-- AT&T currently has 11,000 route miles of fiber optic cable in place. Planned deployment by 1989 is for 24,000 route miles as part of a 53,000-mile digital network. AT&T's 1.7 Gbps fiber system provides inter-city capacity for 24,000 simultaneous calls.

-- MCI began significant fiber optic installation in 1983 and expected to have 19,000 route miles of fiber optic cable deployed by the end of 1988.

-- At the start of 1988, U.S. Sprint had completed 17,000 miles of a planned 23,000-mile fiber-optic network, representing a $2 billion investment. U.S. Sprint anticipated that by the end of 1988 95% of its traffic would be transmitted by fiber-optic cable.

-- Five regional exchange carriers, Williams Telecommunications, Southernet, Litel, Microtel, and Consolidated Network, have formed a consortium, the National Telecommunications Network (NTN), and have installed a 10,500-mile, $1 billion fiber network.

-- 5,000 miles of fiber optic cable are managed by Lightnet, jointly owned by CSX Corporation and Southern New England Telephone Company.

The advance of fiber optic cable is primarily due to its low cost, promise for still higher capacity, and its already mentioned technical advantages. The original Washington, D.C.-New York City route used multi-mode fiber transmitting at 90 Mbps, with repeater spacings of less than 10 km. The high-capacity long-haul routes being installed today typically operate in the 400-800 Mbps range, and sometimes even in the 1 to 2 bps range, with repeater spacings of 30-40 km.[54] TAT-8, a submarine fiber optic cable across the Atlantic, operating since 1988, was designed for 60 to 65 km spacing between repeaters. By the mid-1990s, it is expected that fiber systems will be composed of metal fluorides rather than today's silica base and will carry a gigabyte per second across 3,600 km without repeaters.[55]

The unit cost of fiber optic cable has undergone similarly impressive reductions. From 1981 to 1984 the cost of installing a fiber circuit and its associated electronics dropped by a factor of ten. In 1986, fiber optic cable was being produced at a cost of 30 cents a meter.[56] Estimates for the initial cost of metal fluoride fiber optic cable vary between $100 and $200 a meter, but this may go down to $3.00 to 90 cents a meter. The reduced repeater expense associated with the metal fluoride cable is expected to compensate for its higher unit cost.[57]

Fiber optic cable is being installed instead of coaxial cable or microwave repeaters whenever there is sufficient traffic, existing or projected, to warrant its use. This includes completely new installations as well as the replacement or upgrading of existing systems. Today, fiber optic cable is the medium of choice for high-capacity long-haul, metropolitan inter-office, and subscriber loop feeder applications.

Annual fiber optic cable investment in the U.S. is expected to reach $2.9 billion by 1992, compared to $774 million in 1986. The same figure for Europe in 1991 should be $1.5 billion, given the projected annual growth rate of 80%. Between 1985 and 1990, it is expected that Japan will see an average annual growth in fiber optic demand of 200%.[58]

Even the most staunch supporters of satellites acknowledge that fiber optics offers a technically and economically competitive, if not superior, alternative to satellites in the provision of high capacity, point-to-point voice services such as along the heavy North Atlantic route between Europe and the United States, where the TAT-8 cable has begun service. Furthermore, it can be expected that fiber optics will continue to enjoy rapid technological gains in the next decade that will strengthen its competitive position in terms of performance and cost.

Complementarity of Satellites and Fiber Optic Cable

Straightforward cost comparisons between fiber optic cable and satellites are complicated by the inherent differences between the two transmission media. Real-world cost comparisons must be made on a case-by-case basis. However, it is instructive at this point to call attention to some of the variables involved.

First, it must be recognized that the cost of a fiber optic communications network is distance-sensitive, whereas the cost of a satellite system is not. In constructing a communications network with fiber, there is a cost involved in connecting each node to the network as well as the cost of running cable to the node. This second cost is proportional to the node's distance from the nearest network interconnect point. Implementing new nodes in a satellite communications network requires only the installation of an earth station at the desired site. As long as a site is within the footprint of the satellite, it achieves full interconnectivity with all other nodes in the network and is unaffected by its distance from the rest of the network.

Although owners of fiber optic cable systems need not worry about obtaining the rights to use radio frequencies and positions in the geostationary orbit, rights-of-way must be procured along the route of the cable.

Another consideration in the selection of a communications system is its reliability and longevity. The INTELSAT family of communication satellites has demonstrated a long-term reliability factor of better than 99.9% uninterrupted service. While figures for transoceanic fiber optic systems are not yet available, one must be aware of the difficulty of repairing a cable lying on the ocean floor. A single cable sheath handling thousands of telephone calls simultaneously is highly vulnerable to service interruption caused by an inadvertent or deliberate break in the cable. One safeguard designed to prevent this possibility is to provide a passive branching system in the cable that allows multiple landing points. The first generation of transoceanic fiber optic systems is designed for a useful life of twenty-five years and should not require ship repairs more than three times during that period, a mean time between repairs of about eight years.[59] Since the useful lives of orbiting and planned satellites vary from four to thirteen years, and because the difficulty of repairing a fiber system depends in part on the terrain it crosses, whether it is terrestrial or the surface of the ocean floor, it is difficult to make a general comparison of the longevity and reliability of satellite and fiber optic cable systems.

Traffic intensity, variability, and distribution must also be considered when measuring the relative advantages of fiber optic cable and satellites. In order for fiber to provide a viable economic alternative, it must be operated at or near its stated capacity; fiber's optimum application is thus in high-volume, point-to-point situations. A fiber optic cable providing service in an area bypassed by changing traffic patterns, or where growth has not kept pace with projections, may not be able to provide cost-effective service. In contrast, satellite networks are easily reconfigured to alternate between thin-route and heavy-volume applications, thereby remaining economically competitive in both.

Finally, a variable that further complicates cost projections for fiber optic systems stems from the rapid pace of technological improvements in such systems. Fiber optic systems capable of data rates of several gigabits per second have been demonstrated in the laboratory. Extremely low loss cable systems, on the order of 0.001 dB/km, are theoretically possible. This low attenuation, together with single-mode fibers operating at wavelengths of two to four microns, will allow greater spacings between repeaters with resulting savings in costs. Improvements and innovations in cable splicing techniques are also expected to yield significant cost savings.

It is popular today to think of satellites and fiber optics as being purely competitive forces, going head-to-head in winner-take-all campaigns for the most profitable telecommunications markets. But satellites and fiber optics are vastly different technologies, each with its own inherent advantages that should influence decisions on which technology represents the optimal solution for any given application. In fact, it can be generally stated that in any application for which one technology is ideally suited, the other is not.

What needs to be emphasized and understood by network planners and telecommunication users is that more than being strictly competitive forces, satellites and fiber optics can play complementary roles in the global telecommunications network of the future. This, in fact, is the approach that many of the world's major users are taking.

ROLES FOR SATELLITES IN THE
EVOLVING GLOBAL SYSTEM

In the short span of just three years the tone of the discussion surrounding fiber optics and satellites has undergone an interesting metamorphosis. The first stage was disbelief that satellite applications could be affected by fiber optics. The second stage was marked by a fear that an intermodal showdown was imminent and the future of satellites was dark. The third and current stage recognizes that transitions will occur, but that each medium has its own ideal applications.[60] Increases in capacity and decreases in cost over the past several years have made long-haul fiber optic networks a reality, and additional gains in capacity with still lower costs are expected. In relation to satellite systems, fiber optics will represent direct and formidable competition in certain applications such as international voice and data traffic in the U.S.-to-Europe corridor. But in many cases, the two media will complement each other in hybrid networks where each is used to its best advantage. Individually and collectively, fiber optic cable and satellites represent the primary transmission technologies which will allow the visions of high data rate integrated digital networks to become a reality.

Several sections of this chapter have described the technical and operational strengths of satellites that are presently being exploited to meet the needs of telecommunications users. Trends and developments in satellite communication technology, along with the major markets and applications associated with satellites, were also identified. These elements will be used to sketch a picture of the likely roles that satellites can be expected to play in the evolving world telecommunications system.

Developing Countries

From a global perspective, the need to establish a basic telecommunications infrastructure is increasingly being recognized as a fundamental element in the economic development of third-world countries. Most of the developing countries of the world recognize and utilize the strengths and advantages that satellites offer in providing basic communications services. Through INTELSAT, a number of services including domestic services, Vista, low cost earth stations, SHARE (Satellites for Health and Rural Education), and IADP (INTELSAT Assistance and Development Program), offer either free or leased use of INTELSAT transponders and technical support to developing countries to help achieve social and economic goals.[61] In developing and remote areas, satellites aid economic development by providing a basic telecommunications infrastructure as well as furthering the distribution of health services and educational programming.

Mobile Communications

In areas where users may be out of the reach of public telecommunications facilities, such as aboard ships and aircraft, and in remote areas of the United States and in much of Canada, satellites will enable users, through the use of small, portable terminals, to communicate with each other or to interconnect with the public switched telephone network. User interest and acceptance, government support, and the development of appropriate earth station technologies are bringing mobile satellite services to fruition.

Mobile satellite service is being offered by INMARSAT standard C terminals today. Other forms of commercial services are expected to begin commercial operation in the United States in 1992 and in Canada by 1990. Launches for satellites providing radio determination satellite service have been scheduled for every year from 1987 through 1994. Here satellites have a clear edge. Plugging a fiber optic cable into a car, truck, bus, train, airplane, or boat is not easily accomplished.

Terrestrially-based, cellular communications systems also provide mobile services. Their economics are suited best to high-volume applications, generally urban centers with high densities of mobile users. Some see the volume of mobile service rivaling that of PSTN telephone service by the 21st century. It remains to be seen whether operators of mobile satellites, radio determination satellites, and cellular systems will compete with or complement one another.[62]

Private Data Networks

To meet the ever-expanding needs of business communications, there is a growing trend toward the construction of private data networks. Private data networks have the economic attraction of giving their users direct access to private facilities, bypassing the public switched telecommunications network and avoiding its charges.

Technical capabilities for earth and space-segment facilities have been developed to meet market trends and satisfy user needs. The application of intelligent, microprocessor-controlled, small earth stations facilitates the implementation of flexible, user defined, distributed processing networks. Data networks using very small aperture terminals allow users to access the satellite either directly or through a large, centrally located earth station that acts as a hub and also provides network management functions. Since VSATs are portable and can be used wherever there is direct line of sight access to a satellite, nodes can be easily set up or taken down depending on the needs of the user.   U.S. regulations on the use of VSATs, especially for reception only, are much more relaxed than they are in other countries.

Wideband and Point-to-Multipoint Distribution Services

Because satellites can transmit directly to user sites, they offer a true end-to-end wideband service. Although fiber optics and other transmission media such as coaxial cable also offer wide bandwidth, fiber is primarily employed in point-to-point trunking applications, *i.e.*, between telephone company switching centers.   Until wideband facilities are extended the "last mile" to the user site, the full benefits of fiber's wide bandwidth will not be realized.

In point-to-multipoint applications satellites again have the edge over other technologies. Any user located in the footprint of the satellite can receive the satellite's transmissions. The cost of building the network is not affected by the distance between the nodes in the network. A new node can be established simply by setting up an earth station.

Telephony Services for Residential Customers

The shift in cost and complexity in a satellite network away from the earth station to the satellite is bringing direct access to satellites within the reach of many small users. The small user can also bypass the local exchange carrier and avoid any "last mile" costs or access charges.

Long-Distance and Transoceanic Voice and Data Traffic

Although the new submarine fiber optic cables are expected to capture a significant portion of the transoceanic market, satellites will continue to provide a large portion of these services in the foreseeable future, particularly for developing countries and thin route communications. Their great flexibility allows satellites to provide cable restoration services during periods of cable outages and repair.

34

NOTES

1. ITU, CCIR, *Technical Bases for the World Administrative Radio Conference on the Use of the Geostationary-Satellite Orbit and the Planning of the Space Services Utilizing it (WARC-ORB (1)*, Geneva, 1984, Part II, pp. 32-71.

2. W.J. Howell, *World Broadcasting in the Age of the Satellite: Comparative Systems, Policies, and Issues* (Norwood, N.J.: Ablex Publishing Corp., 1986), pp. 247-263.

3. See D. Athanassiadis, "The Canadian MSAT Program," *MSAT-X Quarterly* Feb. 1986, pp. 3-4.

4. AMSC is an eight member consortium made up of Hughes Communications Mobile Satellite Services, Inc., McCaw Space Technologies, Inc., Mobile Satellite Corp., Satellite Mobile Telephone Co., Skylink Corp., Transit Communications, Inc., and North American Mobile Satellite, Inc.

5. See L.C. Steele, "MSS Update," *MSAT-X Quarterly*, Oct., 1989, p. 1.

6. M. Rothblatt, "Radiodetermination Satellite Service," *Telecommunications*, June, 1987, pp. 39-42.

7. "Satellite Industry Worth $500 Million," *Satellite Communications*, Oct., 1986, p. 15.

8. "FCC OKs construction, launch, of 9 new COMSATs, 10 replacements, for 42 total in orbit," *Telecommunications Reports*, Nov. 21, 1988, p. 8.

9. See the section of this chapter beginning on page 19 for an overview of VSATs.

10. Tamara Bennett, "Private Networks 1988," *Satellite Communications*, July 1988, pp. 32-39.

11. Wilbur Pritchard, "The Basics of Satellite Technology," in Pelton and Howkins, p. 20.

12. The higher frequency is used for the uplink so that a lower frequency can be used for the downlink. This helps to minimize the problem of a weak incoming signal being distorted and fading due to rain attenuation. This problem becomes worse as the frequency increases.

13. Pritchard, pp. 19-20.

14. Mark Long, *World Satellite Almanac* (Indianapolis: Howard W. Sams and Co., 1987), pp. 40-41.

15. Joseph N. Pelton, "A User-Friendly Introduction to Satellites," in Pelton and Howkins, pp. 2-3.

16. F. Michael Naderi and S. Joseph Campanella, *NASA's Advanced Communications Technology Satellite (ACTS): An Overview of the Satellite, the Network, and the Underlying Technologies*, NASA, March 1, 1988, p. 1, 4; and University of Colorado Center for Space and Geosciences Policy, *Report of the ACTS/Science Workshop: Potential Uses of the Advanced Communications Satellite to Serve Certain Communication, Information, and Data Needs to the Science Community* (Boulder, CO: University of Colorado, Center for Space and Geosciences Policy), 1988, second page of Attachment.

17. Naderi and Campanella, p. 2.

18. Charles W. Bostian and Timothy Pratt, *Satellite Communications* (New York, NY: John Wiley and Sons, 1986), p. 69.

19. Naderi and Campanella, p. 3.

20. Bostian and Pratt, p. 235.

21. University of Colorado, Center for Space and Geosciences Policy, Attachment B.

22. Joseph N. Pelton and William W. Wu, "The Challenge of 21st Century Satellite Communications: INTELSAT Enters the Next Millennium," *IEEE Journal on Selected Areas in Communications*, May 1987, pp. 585-587.

23. Naderi and Campanella, pp. 3-4.

24. *Ibid.*, pp. 13-14.

25. *Ibid.*, p. 7.

26. Jitter due to high winds can create loss of satellite signals.

27. "Proof of Concept for ACTS Technology Started in 1980," *ACTS Update*, NASA: Lewis Research Center, Issue 86/4, Oct., 1986.

28. "Lasercom: Pioneer in Intersatellite Links," *ACTS Update*, NASA: Lewis Research Center, Issue 86/2, April, 1986.

29. Naderi and Campanella, pp. 13-14.

30. INTELSAT, *INTELSAT Report 1987-88*, Washington, D.C., 1988, pp. 3 & 7.

31. *Ibid.*, p. 7.

32. Dean Burch, "INTELSAT: the Tomorrow Organization," in Pelton and Howkins, p. 26.

33. INTELSAT, *INTELSAT Report, 1986-87*, Washington, D.C., 1987, p. 20.

34. Edwin D. Parker, "Micro Earth Stations as Personal Computer Accessories," *Proceedings of the IEEE*, Nov. 1984, pp. 1529-1530.

35. Higher rates are possible with more expensive and larger antennas.

36. Scott Chase, "Teleports: What they are depends on who's talking," *Via Satellite*, Aug. 1987, pp. 22-25.

37. Gerhard Hanneman, "Teleports: The Global Outlook," *Satellite Communications*, May 1987, pp. 29-33.

38. Chase, pp. 23-24.

39. Bostian and Pratt, p. 415.

40. *Ibid.*, pp. 309-312.

41. *Ibid.*, p. 415.

42. *Ibid.*, pp. 281-284.

43. P.R.K. Chetty, *Satellite Technology and its Applications* (Blue Ridge Summit, PA: Tab Books, Inc., 1988), p. 317.

44. Bostian and Pratt, p. 261.

45. *INTELSAT Report 1986-87*, p. 36.

46. See Annex 3 for the OSI model.

47. P.J. McDougal and J.N. Pelton, "ISDN: the Case for Satellites," *Telecommunication Journal*, vol. 54, No. V (May 1987), pp. 318-322.

48. H. Charles Baker, John C. Bellamy, John L. Fike, and George E. Friend, *Understanding Data Communications* (Indianapolis, IN: Howard W. Sams & Co.), 1984, pp. 141-143.

49. Uyless Black, *Data Communications and Distributed Networks* (Englewood Cliffs, N.J.: Prentice Hall, Inc., 1987), pp. 55-60.

50. Fiber optic cables, however, are vulnerable to sharks, trawlers, and sabotage.

51. Lawrence A. McLernon, "Fiber Optics as a Replacement Technology," in *Conference Proceedings, Fibersat 86* (Vancouver, Canada: Wescom Communications Studies and Research Ltd., 1986), p. 347.

52. Michael Warr, "Kessler Projects FO Expansion," *Telephony*, June 27, 1988, pp. 13-14.

53. Tom Valovic, "Fiber-Optic Deployment Among the Interexchange Carriers," *Telecommunications*, May 1987, pp. 40-53; and U.S., Dept. of Commerce, National Telecommunications and Information Administration, *NTIA Telecom 2000*, Washington, D.C. 1988, p. 260.

54. See David Kalish and Fred Feldman, "Fiber Optic Trends in the United States," *Conference Proceedings, Fibersat 86*, pp. 331-332.

55. Paul Strauss, "Lightwave Future Gets Even Brighter," *Data Communications*, Feb. 1987, p. 53.

56. See McLernon, p. 345-349 and R.A. Gershon, "Satellites and Fiber Optics: Redefining the Business of Long-Haul Transmission," *Business Communication Review*, Vol. 18, No. 2, March-April 1987, pp. 22-25.

57. Strauss, p. 56.

58. *NTIA Telecom 2000*, p. 266.

59. TAT 8, the first transatlantic fiber optic cable has already experienced several failures necessitating the repair of each of its major segments.

60. Anthony M. Rutkowski, "Beyond Fiber Optics vs. Satellites," *Telecommunications*, Sept. 1985, pp. 11-15.

61. *INTELSAT Report, 1986-87*, p. 34.

62. "Industry Experts Pragmatic About Satellites' Role in Land Mobile Services," *Satellite Communications*, June 1988, p. 8.

CHAPTER TWO

U.S. POLICY OF GLOBAL MONOPOLY

AND ITS AFTERMATH

INTRODUCTION

The tremendous lead in satellite communication technology achieved by the U.S. in response to the shock of Sputnik permitted America's leaders for many years to decide just how the technology and its use should evolve. Early American satellite communications policy was sparked by foreign policy considerations as well as a genuine desire to bring the advantages of satellite communication to the people of the world. The first manifestation of that policy was the decision to create an international commercial satellite entity that would exercise a virtual monopoly over long-distance communications. It would be open to the participation of all countries, and the United States would provide it with the technology necessary for it to succeed. The International Telecommunications Satellite Organization (INTELSAT) was to become that entity.

The potential benefits of satellite communication over conventional methods of long-distance communication were so attractive, however, that in just two decades INTELSAT's monopoly was being increasingly challenged as a result of the creation of new satellite systems both international and regional.

This chapter will detail the origins of the U.S. policy of global monopoly, the manner in which it was implemented, and the growing network of long-distance satellite communication systems that was soon to follow.

SPUTNIK AND ITS IMPACT

The initial U.S. response to the flight of the Soviet Union's Sputnik in 1957 was one of shock and dismay at the fact that the USSR had evidently taken the lead in the use of an important new technology. However, once the initial shock had worn off, the United States quickly began to pursue two interrelated objectives: 1) to catch up to and surpass the USSR in terms of space technology; and 2) to use this new technology to score a diplomatic triumph over the USSR.

The first objective was attained with the launching in 1958 of Explorer, a scientific satellite. Soon after there followed a series of communications ' satellites named Score, Courier IB, and then Telstar I. The latter, built by Bell Telephone Laboratories, was the world's first active repeater satellite. Telstar I, placed into an elliptical orbit on July 10, 1962, transmitted on the 4 GHz band and received on the 6 GHz band. The success of Telstar I experiments, along with other active communication satellites such as Relay and Syncom and various Department of Defense projects, paved the way for achieving the second objective, regaining U.S. prestige lost with the launching of Sputnik in 1957.[1] Almost from the beginning of its efforts to catch up to the U.S.S.R., the U.S. took pains to emphasize that U.S. satellites would be used to create a worldwide communication system for the benefit of all nations. This theme was expressed by President Eisenhower just before leaving office and repeated by President John F. Kennedy in a statement issued on July 24, 1961. In that statement, in which he asked Congress for additional funds for the effort, he stated: "I . . . invite all nations to participate in a communication satellite system in the interests of world peace and closer brotherhood among peoples throughout the world."[2]

COMSAT

The United States then had to confront the issue of the form that U.S. participation should take in this yet to be organized international entity which, it was believed, would become dominant in the field of international telecommunications. Some senators, led by Estes Kefauver, were of the opinion that because of the heavy investment of the government in the R&D of communication satellite technology, the government itself should be the U.S. partner in this new international entity. However, the majority, led by Senator Robert S. Kerr, felt that in the spirit of the U.S. tradition of free enterprise the U.S. partner should be a private entity.

After extensive consultation with interested government agencies and those portions of the private sector that might in any way be affected by the new entity, the U.S. Congress passed the Communications Satellite Act of 1962 (by 353 to 9 in the House and 66 to 11 in the Senate). This act created COMSAT, a private corporation with government oversight to be the sole authorized U.S. operator of international satellite communications facilities and the U.S. participant in the new international satellite communications system.[3]

COMSAT was to be financed through the sale of shares, 50% reserved for authorized international common carriers and the remaining 50% set aside for the general public and other private entities. COMSAT was to be run by a 15-member Board of Directors, six of whom were to represent American international carriers, six to represent the general public shareholders, and three to be appointed by the President with the advice and consent of the Senate.[4]

Government supervision, which was to be extensive, included, in addition to the appointment of three of the members of the Board of Directors by the President, a requirement to report the activities of COMSAT to the government, NASA's responsibility for providing R&D, and, on a reimbursable basis, satellite launching and associated services. COMSAT was also subject to oversight activities by the Federal Communications Commission, including the approval of rates, stock offerings, and new services; ensuring that COMSAT's services are open to all without discrimination; enforcing competitive bidding on hardware; and granting construction permits.[5]

INTELSAT

Negotiations with Allies

The policy of the United States concerning the new international satellite communications system is set forth in Section 102 of the Communications Satellite Act of 1962: 1) it would be universal; 2) it would be a commercial enterprise; 3) it would be open to participation by any country in the world; and 4) it would be used to provide services to both developed and developing countries.

Sec. 102 reads:

(a) The Congress hereby declares that it is the policy of the United States to establish, in conjunction and in cooperation with other countries, as expeditiously as practicable, a commercial communications satellite system, as part of an improved global communications network, which will be responsive to public needs and national objectives, which will serve the communication needs of the United States and other countries, and which will contribute to world peace and understanding.

(b) The new and expanded telecommunication services are to be made available as promptly as possible and are to be extended to provide global coverage at the earliest possible date. In effectuating this program, care and attention will be directed towards providing such services to economically less developed countries and areas as well as those more highly developed, toward efficient and economical use of the electromagnetic frequency spectrum, and toward the reflection of the benefits of this new technology in both quality of services and charges for such services.

COMSAT, along with representatives from the U.S. government, began almost immediately to translate this directive into reality. Discussions were begun with Canada and the United Kingdom, followed by the Conference of European Postal and Telecommunications Administrations (CEPT) in 1963. Most of the major problems that divided the U.S. and its friends, including the decision not to adopt a definitive agreement until later, were ironed out in a meeting between the U.S. and CEPT in Rome in 1964 at a time when only the U.S. among the negotiators had the technology necessary for such an endeavor.

The end product was a treaty entitled "Agreement Establishing Interim Agreements for a Global Communications Satellite System," signed by representatives of fourteen interested governments and the Vatican City on August 20, 1964. The parties agreed to meet again in 1969 to draft a definitive agreement. Until that time the new international organization was to be composed of an Interim Communications Satellite Committee made up of the signatories to provide overall policy control and COMSAT, which was to manage the system, including design, construction, operation and maintenance. COMSAT was given 61% ownership of the interim system and the others 39%, based on projected international and domestic use. A total of 17% of the latter portion was reserved for new members.

Voting in the interim system was based on ownership. Important questions were to be decided by a three-quarters majority. Non-substantive issues were to be resolved by a simple majority. As it turned out, most decisions were made without resorting to a vote, a tradition of consensus that remains very much alive to this day.

The parties to the Interim Agreement, which had grown in five years to eighty-four, began meeting on February 24, 1969, to draft a Definitive Agreement. By August 20, 1971, two and a half years later, they had drafted and signed a Definitive Agreement which came into force on February 12, 1973. Three major problems of interest to us here were solved at these meetings between the United States and its major INTELSAT partners: 1) who should manage the system; 2) whether there should be a single universal system or many systems; and 3) whether or not the entity should be restricted in the types of services it would provide.

With respect to system management it was acknowledged by most who participated in the meeting that only COMSAT had the technological expertise to actually run the system; but at the same time others felt that this knowledge would not always remain in American hands, and when a diffusion of that knowledge occurred, others should be allowed to participate. The solution, spearheaded by Japan and Australia, was to divide the functions of INTELSAT into two parts, one under a Secretary General with an international staff to handle administrative, financial and legal responsibilities, and the other to be headed by a Management Service Contractor (i.e. COMSAT) to handle technical and operational responsibilities. The first would be carried out as soon as possible and the second in six years' time. During those six years, COMSAT would be responsible for technical and operational functions.[6]

In the discussions regarding whether additional international satellite systems would be permitted, it is interesting in light of later events that the United States argued strongly that it was necessary to have a single universal system if INTELSAT were to be economically viable. Others argued that there were many reasons why it would be necessary to organize additional systems and that members of INTELSAT should not be forbidden to do so. The solution to this problem was also a compromise. While the Preamble and various articles in the Definitive Agreement assert in almost unequivocal terms that a single system should be the aim of all of those who were to participate in INTELSAT, an article was added outlining the procedure for the creation of separate systems. The party or parties contemplating the creation of a separate system must notify that fact to INTELSAT's Assembly of Parties through the Board of Governors and must provide all relevant information concerning the new system. The application is subject to review under Article XIV of the INTELSAT Agreement for: (1) "technical compatibility" with the use of the radio frequency spectrum and orbital space; (2) whether or not the new system would cause "significant economic harm" to INTELSAT; and (3) whether or not the system would "prejudice the establishment of direct telecommunication links through the INTELSAT space segment among all participants."[7] However, the decisions of the Assembly of Parties on these concerns take the form of a non-binding recommendation![8]

On the third issue, that of restricted or unlimited services, a compromise was reached whereby specialized services would be permitted, but the decision to do so was made a substantive issue requiring a two-thirds majority for approval.[9]

INTELSAT's Structure

The final form of INTELSAT's structure was also a compromise, but this time between the interests of the developed nations and those of some of the developing nations. The developing countries wanted INTELSAT's major policy-making organ to take a form similar to that of the General Assembly of the United Nations, in which each member country had the right to be represented and each had one vote. The United States and its supporters, however, wanted power to be concentrated in a smaller body with voting power commensurate with investment in the system.

To meet the objectives of the two groups, INTELSAT's eventual structure was comprised of three major components, the Assembly of Parties, the Meeting of Signatories, and the Board of Governors. The Assembly of Parties is made up of member states, each with one vote, and meets every two years to discuss overall policy. It has the power to recommend changes to the basic treaty that should be submitted to member states. Such changes come into force if agreed to by two thirds of the signatory states and if that two thirds includes states with two thirds of the overall investment in INTELSAT or 85% of the membership.[10]

The Meeting of Signatories is composed of states or designated representatives, COMSAT in the case of the United States, and meets once a year. The Meeting of Signatories approves recommendations of the Board of Directors in such matters as: a) approval of earth station access to the INTELSAT space segment; b) allotment of INTELSAT space segment capacity; c) establishment and adjustment of the rates for the use of INTELSAT space capacity; and d) the investment shares of participants. Each participant has one vote. Procedural matters are decided by majority vote but a two-thirds majority is necessary on matters of substance.[11] INTELSAT also has an Executive Organ, headed by a Secretary-General, located in Washington, D.C.

The Board of Governors, which manages the system, is made up of one governor for each country with more than the minimum share established by the Meeting of Signatories, one governor for each of two or more signatories whose combined investment share is over the same minimum, and one governor for each group of at least five signatories not represented in the other categories from any one of the regions defined by the ITU.[12]

COMSAT Operations Within INTELSAT

In addition to controlling the appointment of three of COMSAT's fifteen directors, the U.S. Executive Branch determines how COMSAT will vote at INTELSAT meetings. The State Department's Bureau of Communications and Information Policy, working with the Commerce Department and the Federal Communications Commission, establishes the COMSAT position.

U.S. international carriers usually gain access to satellite service through COMSAT, INTELSAT, and a foreign INTELSAT representative, typically a government-owned Post, Telegraph, and Telecommunications (PTT) Ministry. U.S. carriers lease INTELSAT capacity from COMSAT in units of half-circuits. COMSAT then connects the transmissions of its clients to half-circuits leased by the other country's INTELSAT representative, which receives revenues for the use of its half-circuit.[13]

INTELSAT'S ACCOMPLISHMENTS

INTELSAT certainly lived up to the expectations of its founders. The world's first and largest commercial satellite system, INTELSAT has been the dominant carrier of international communications since 1964, handling more than two thirds of all transoceanic telecommunications. In 1989 INTELSAT relayed more than a billion telephone conversations.[14] INTELSAT currently manages the operation of thirteen satellites and 767 earth stations. Membership in the organization increased from ninety-four in 1976 to 102 in 1979 and to 117 in 1989. The total number of countries, territories, or possessions served by INTELSAT (directly or indirectly) increased from fifteen in 1966 to 172 in 1988; and the number of channels in service increased from 150 in 1965 to 120,000 in 1989. Net investment in INTELSAT rose from $69,263,000 in 1967 to $1,512,000,000 in 1987 and total revenues from $143 million in 1976 to $519 million in 1987. All of this occurred while the annual space segment utilization charge (per half circuit) dropped from $32,000 per annum to $4,440.[15] With the introduction of new circuit multiplication equipment, these rates are expected to drop to about $1,000 per annum.

One of the keys to INTELSAT's success has been the introduction of new services. In September, 1983, for example, INTELSAT announced a new service called INTELSAT Business Service (IBS), a fully digital service specifically designed for business needs. IBS services include fully digital applications for high and low speed facsimile and data, video, teleconferencing, voice, electronic mail, and telex. These services are leased for part-time or full-time purposes and may transmit from 64 kilobits per second to 8.44 megabits per second or higher if requested. Even though the system has close to global connectivity, IBS was designed for single hop, point-to-multipoint transmissions, in the C or K band frequency spectrums. It is especially designed to connect the West Coast of North America and Europe to locations in the Middle East. IBS cannot be accessed via the public switched telecommunications network (PSTN), but INTELSAT offers three options for its access:[16]

1) user gateways which permit businesses to have direct access through small on-premise earth stations (Standard E-1 and F-2 antennas);

2) urban gateways which allow groups of users in a given location to share medium-sized earth stations (Standard E-2 and F-2 antennas); and

3) country gateways which route services through large earth stations currently in place to receive present INTELSAT transmissions (Standard A,B,C, F-3 and E-3 antennas).

INTELSAT has established six IBS earth station standards, as identified above, in addition to the organization's Standard A, B, and C earth stations. The diameters of the six stations range from the E-1 at 3.5 meters to the F-3 at 9 meters. The appropriate choice of the three options depends on the traffic volume which an earth station must support.[17]

The use of IBS has enjoyed spectacular growth, expanding from 14 full-time, 64 kbps channels in 1984 to 3,646 channels in 1987. The 1987 figure represents a 134% increase over 1986. Two grades of service quality are offered, Basic IBS (in C or K band) and Super IBS (in K band). In the higher power C and K band versions bit error rates of 10^{-8} are normally provided with 10^{-3} representing the most degraded conditions. IBS, according to INTELSAT, is of "ISDN quality."[18] INTELSAT offers IBS on full-time, part-time, or occasional use basis. Customers can also use it via leased, full-time transponders. Leases in multiples of 9 MHz of bandwidth are available for periods of one, four, or seven years.

In December, 1985, INTELSAT introduced Planned Domestic Service (PDS), which allows member countries to purchase transponders for domestic television broadcasting and other domestic services. This program allows signatories to obtain more direct control of a domestic satellite service, and has brought $112.73 million of additional funds to INTELSAT. This capital influx will lower the organization's overall revenue requirement without compromising its overall quality of service, since the transponders being sold are excess capacity C and Ku band transponders.[19] Signatories may also lease transponders for domestic service on either a preemptible basis or a "planned" basis. Almost thirty countries rely on INTELSAT for domestic long-distance and rural services, a number which INTELSAT believes could double by the end of the 1990s.[20] Table 2.1 shows 1988 rates for Planned Domestic Service and transponder leases.

Table 2.1

**Rates for Planned Domestic Service and
Transponder Leases, 1988**

Transponder Type (Bandwidth in MHz)	Sale Price (millions US $)	Annual Lease Price (millions US $)
C-Band Hemi (72)	9.400	1.777
C-Band Hemi (36)	6.435	1.215
C-Band Zone (72)	9.400	1.777
C-Band Zone (36)	6.435	1.215
K-Band Spot (72)	14.053	2.676
K-Band Spot (150)	26.498	5.049

Note: These rates are reduced for satellite transponders with less than a full projected lifetime.

Source: INTELSAT Annual Report by the Board of Governors (1988), p. 11.

The total number of Time Division Multiple Access (TDMA) half-circuits in the INTELSAT system reached 6,144 in 1987. Of that total, 3,748 half-circuits, or 61%, served the Atlantic region. The countries with TDMA terminals connected to INTELSAT TDMA services were: Australia, Belgium, Canada, France, Germany (FR), Hong Kong, India, Indonesia, Italy, Japan, Korea (Republic of), the Netherlands, the Philippines, Singapore, South Africa, Spain, Sweden, Switzerland, the United Arab Emirates, the U.K, and the U.S.[21] During 1987, the INTELSAT space segment provided 524,238 channel days of occasional-use channel services. This included 430,852 channel days on thirty-five different occasions for back-up services to other transoceanic or terrestrial telecommunications systems.[22]

INTELSAT occasional-use television services accounted for 55,778 hours of channel traffic in 1987, averaging 108 television transmissions per day.[23]

In the early 1980s, INTELSAT began the first full-time international satellite television service, starting with broadcasts between the U.S. and Australia.[24] Channels dedicated to full-time television transmission may be leased for long-term, short-term, or part-time periods, on a preemptible or non-preemptible basis. Long-term leases are for two, five, or seven years, while short-term leases have a minimum duration of three months. Part-time leases stipulate at least once-a-day usage during one year. At the end of 1987, the number of full-time international television leases totaled 27, all but four of which were for long-term use.[25]

INTELSAT has noted the growth in worldwide demand for international video services, including those received via smaller earth stations accessed by cable TV systems, as well as those received via small TV receive-only earth terminals. Interest in satellite news gathering services also grew in 1987, and INTELSAT is looking for new ways to provide such services. Recently approved television services include digital video transmission.[26]

INTELSAT broadcast services also include radio services on analog or digital carriers, both of which are restricted to one-way transmission.[27]

1987 saw the introduction of Intelnet service, a point-to-multipoint offering which permits the implementation of VSAT technology in business applications. Large INTELSAT earth stations, the Standard A, B, C, and E-3, function as central nodes, connected to VSATs in star topologies. Data distribution or non-interactive service is available through the Intelnet I offering, while the Intelnet II service (interactive) enables data collection from remote locations as well as data distribution. VSATs with antennas of 0.6 to 2.5 meters are compatible with the service, so long as they meet performance standards for INTELSAT's Standard G earth stations. Intelnet service is available for international and domestic use and is leased in 1 MHz increments in the K band and 4.5 MHz segments in the C band. Users may elect preemptible or non-preemptible service. The first such commercial lease was that between the U.S. and Argentina. Intelnet connections for the U.K., Hong Kong, Singapore and Australia are also in operation or have been approved.[28]

With a primary goal of providing Caribbean countries with digital telecommunications services on a cost-effective basis, INTELSAT's Caribnet service offers them IBS and Intelnet services at a 50% discount. The service is limited to the intra-Caribbean traffic of Caribbean signatories and non-signatories.[29]

Bolivia, Equatorial Guinea, Maritius, Vietnam, and some countries in the South Pacific are among those that benefit from INTELSAT's Vista and Super Vista services. These new services were established to provide voice and low-speed data capabilities to and among isolated or remote areas with light international or domestic traffic loads. Vista service is accessed in units of one channel while Super Vista, accessed on a DAMA basis, is billed on a channel unit access basis. At the end of 1987, 128 Vista channels were in service, compared to 50 in 1986. The Vista service allocation on transponder 37 of INTELSAT's 63° west satellite reached the saturation point in 1987, and a waiting list is now in effect.[30]

At the end of 1987, 456 intermediate data rate (IDR) carriers were in service. Hong Kong, the U.K., and the U.S. were the sole users. The Indian Ocean region accounted for 406 of the IDR carriers, and the Pacific Ocean region accounted for 50.[31] INTELSAT also offers multidestinational IDR carriers.[32]

During its 1987 meeting the INTELSAT Board of Governors authorized the use of mobile voice and data services on a special and temporary basis for short-term or emergency situations. The authorization did not include mobile video services, which were already available for special events. In addition to emergency situations, the anticipated mobile service clients include heads of state on diplomatic missions. Charges are based on the total pro-rated daily average for the particular INTELSAT service used, plus an earth station rate adjustment factor (RAF) and a 10% surcharge.[33]

In 1984, INTELSAT launched the INTELSAT V-A, which carried 15,000 two-way telephone circuits, in addition to two television channels and two steerable spot beam antennas designed to provide domestic leasing services.[34] The two remaining INTELSAT V-A satellites are still to be launched, and all five INTELSAT VI satellites have been modified to provide higher power at Ku band frequencies. Increased emphasis is being placed on Ku band systems for two main reasons, to eliminate interference problems with C band terrestrial microwave systems, and to offer the ability to utilize smaller earth terminals working to Ku band spot beams.

Tables 2.2 and 2.3 provide an overview of INTELSAT's service to the world community.

OTHER COMMON USER ORGANIZATIONS

This section covers the origins, decision-making structures, and satellite systems of other government-operated common user organizations. The international common user organizations whose membership includes states which are also INTELSAT members, it should be noted, have all argued successfully that their international systems do not violate Article XIV of the INTELSAT Agreement concerning the conditions for approving separate systems.

Table 2.2

INTELSAT Full-Time Satellite Use by Region in 1988

1) Atlantic	69,643
2) Pacific	25,069
3) Indian Ocean	23,700
4) 359° E. Sat	398
Total	118,810

Source: INTELSAT, *Facts & Figures*, March, 1989, p. 10.

Table 2.3

INTELSAT Full-Time Satellite Use by Service in 1988

Service	No. of Channels
FDM/FM	92,395
CFDM	3,726
Single Channel Per Carrier	4,437
TDMA	8,234
INTELSAT Business Service	7,522
Vista	142
Intermediate Data Rate	2,354
Total	118,810

Source: INTELSAT, *Facts and Figures*, March, 1989, p. 9.

INTERSPUTNIK

The INTERSPUTNIK organization was proposed by the USSR in 1968 and the Intergovernmental Agreement on the Establishment of INTERSPUTNIK came into force in 1972. INTERSPUTNIK's eight original signatories founded the organization as a direct response to INTELSAT and its strong American influence. INTERSPUTNIK was given a simplified structure composed of a Board of Members, a Directorate headed by a Director General, and an Auditing Committee. The Board meets at least once a year and is comprised of representatives of the members, each with a single vote. Decisions on substantive issues require a two-thirds majority. The Board is responsible for the operation of INTERSPUTNIK's space segment, specifications for earth stations, rates for leasing channels, and the election of the Director General.[35]

INTERSPUTNIK's Director General is the chief executive officer of the system and is responsible for carrying out the decisions of the Board. The Director General is appointed to a four-year term. INTERSPUTNIK's Directorate has been internationalized and is located in Moscow. The three members of the auditing committee, which oversees INTERSPUTNIK's finances, are appointed by the Board to three-year terms.[36]

The agreement which established INTERSPUTNIK stipulated that the system would evolve in three stages. In the first stage, the USSR would provide free capacity on Soviet satellites while the system's earth segment was being made operational. In stage two, members would lease capacity on Soviet satellites. Finally, in stage three, INTERSPUTNIK's own space segment would be established.

The first global rival to the INTELSAT system, INTERSPUTNIK now consists of fourteen members, Afghanistan, Bulgaria, Cuba, Czechoslovakia, East Germany, Hungary, North Korea, Laos, Mongolia, Poland, Rumania, the USSR, Vietnam, and South Yemen.[37] At least eight non-signatories, including Algeria, Iraq, Libya, Nicaragua, and Syria, utilize INTERSPUTNIK for communication with INTERSPUTNIK members. Prospective members of INTERSPUTNIK need only send a letter of intent to the Directorate agreeing to pay the annual dues and comply with the technical parameters of the system.[38]

Despite rates which are much lower than those of INTELSAT, and a strong promotional campaign directed at developing countries since 1980, INTERSPUTNIK has not been a strong challenger to INTELSAT, which has benefited from quick growth and superior technology.[39] This situation could change. Intercosmos, under East German leadership, is involved in an expanded R & D program, including experiments in the Ku and Ka bands and in the use of small antennas, which could help close the technology gap with INTELSAT.

More important, if INTELSAT is forced to raise its rates because of competition in its more lucrative markets, INTERSPUTNIK's lower rates could prove to be attractive to many of INTELSAT's members, especially the developing countries.[40] Also of possible interest to the developing countries is the policy of one country, one vote in INTERSPUTNIK's Board of Members.

INTERSPUTNIK's former Director General, Spartak Kurilov, has argued for the coordination of INTERSPUTNIK and INTELSAT operations in order to fully exploit each system's unique advantages. He has expressed disappointment with the lengthy process which INTELSAT obliges its members to follow before they can access INTERSPUTNIK even for a short duration. COMSAT vice-president Jack Hannon, however, has noted that COMSAT has proposed that INTELSAT eliminate the coordination process when other satellite systems wish to carry INTERSPUTNIK program material for periods of less than 30 days. In actual practice, the Board of Governors of INTELSAT now has the authority to grant INTELSAT members more frequent use of other systems in certain situations.[41]

INTERSPUTNIK utilizes a leased portion of four Soviet Statsionar (Gorizont) satellites, located in the geostationary orbit at 14° W, 53° E, 80° E, and 140° E, and a network of eighteen earth stations.[42] The leased space segment capacity is jointly owned and operated, with each member state building and owning its own earth stations. All members now own stations, with the exception of North Korea and Rumania.[43]

INTERSPUTNIK provides a variety of voice, data and video transmission services to member nations. All of its satellites operate in the C band, although some also have the capability for Ku band spot beam transmissions.

INMARSAT

The International Maritime Satellite System (INMARSAT) came into being in 1982. Maritime mobile communications had always been affected by atmospheric disturbances, and satellites appeared to be an excellent way to overcome this problem. In 1966 the International Maritime Organization (IMO), then known as the Intergovernmental Maritime Consultative Organization (IMCO), brought together a group of experts to look into the use of satellite communication for maritime purposes. The findings of this group led the IMO to convene the Conference on the Establishment of an International Maritime Satellite System in April 1975, resulting in a Convention and an Operating Agreement. The Convention and Operating Agreement became operative in July 1979, and the INMARSAT maritime satellite system began service on February 1, 1982. At the end of 1987, INMARSAT membership totaled 53.[44]

According to Soviet scholar Jasentuliyana, INMARSAT arose separately from INTELSAT "because important maritime countries were either not INTELSAT members (*e.g.* the USSR), or would have only minor control over maritime satellite policy as a result of a relatively low utilization of the overall INTELSAT system."[45]

INMARSAT has three organs, an Assembly, a Council, and a Directorate headed by a Director General. The Assembly is made up of representatives of all member states and meets every two years. It reviews the activities of INMARSAT, formulates general policy, and makes recommendations to the Council. Each member country has one vote in the Assembly.

INMARSAT's Council functions as a board of directors, responsible for programs and the operation of the space segment. It meets three times a year and is comprised of representatives of the eighteen signatories with the largest investment shares and four others elected on a geographical basis with due regard to the interests of developing countries. The minimum investment share in INMARSAT is 0.05. The top ten signatories in investment shares as of 31 January are listed in Table 2.4.

The Directorate, which is the permanent staff of INMARSAT, is internationalized and located in London. The Director is responsible to the Council, which appoints the Director to a six-year term.[46]

Table 2.4

INMARSAT's Top Ten Shareholders, 1988

Name	Percent
COMSAT	27.5%
British Telecommunications	15.2%
Norwegian Telecommunications Administration	14.0%
Kokusai Deshin Denwa Co. Ltd. (Japan)	9.5%
France Telecom	3.4%
Morsviazsputnik (USSR)	3.3%
Telecommunications Authority of Singapore	2.7%
Hellenic Telecommunications Organization (Greece)	2.7%
Netherlands PTT Administration	2.2%
Compania Telefonica Nacional de Espana (Spain)	2.0%

Source: INMARSAT, *Annual Review 1987/1988*, p. 38.

INMARSAT provides telephone, telex, low/medium speed data over telephone channels, facsimile, leased circuits for voice and high speed data, data communications, limited land mobile applications, group calls, and distress and safety communication services to shipping and offshore communities. Over 7,000 ships use INMARSAT services. The number is expected to rise to 11,000 by 1990 and to reach 17,000 by 1995. In the light of the introduction of the new small, low-cost Standard C Antenna this number may rise even more rapidly. Offshore oil rigs are also users of INMARSAT services.[47]

INMARSAT began service by taking control of satellites previously operated by Marisat. INMARSAT now leases capacity on five European Space Agency MARECS satellites; one is used as a primary over the Atlantic Ocean, one as a primary over the Pacific, and three are used as spares. INMARSAT also leases capacity on four maritime packages located on three INTELSAT satellites, one used as primary for the Indian ocean and the other three as spares.[48]

The Signatories own and operate the system's fourteen coastal earth stations, which provide the interface between INMARSAT's satellites and national and international telecommunication networks on land. INMARSAT's coastal earth stations transmit to satellites on 6 GHz and receive on 4 GHz.[49] Ship antennas are purchased or leased by the ship's owners from manufacturers and suppliers of such equipment. Ship antennas "are typically .85 to 2-meter parabolic antennas housed in a fiberglass radome and mounted on a stabilized platform which enables the antennas to track the satellite despite ship movement."[50]

INMARSAT expects that its mobile satellite communications services will be a $500 million market by 1995.[51]

INMARSAT is also beginning to provide land and aeronautical mobile services. Believing that the system would not be feasible if it were dependent on maritime revenues alone, the organization proposed in 1985 to include the provision of aeronautical services in its charter. This amendment came into force when the number of countries which approved it reached the necessary thirty.[52]

The 1987 ITU Mobile World Administrative Radio Conference decided to assign aeronautic passenger communications secondary status in the Aeronautical Mobile Satellite Service (AMSS) band. An additional outcome of the Mobile Conference was the reallocation of 4 MHz of AMSS band frequency to land mobile service allocation.[53]

The introduction of INMARSAT aeronautical services is being coordinated with relevant international organizations. The INMARSAT Council has approved a draft agreement of cooperation with the International Civil Aviation Organization (ICAO) which delineates functional responsibilities among the two organizations, calls for close cooperation in areas of common interest, and gives each organization reciprocal observer status at the other's meetings.[54] In October 1987, INMARSAT relayed the first satellite telephone call from a scheduled airline flight, a Japanese Airlines Boeing 747 flying over the Pacific Ocean.

Several countries and organizations have experimented with and have plans for aeronautical services via INMARSAT. Singapore, Britain, and the Nordic countries announced their intention to have aeronautical ground station facilities ready for operation in 1989. The same signatories plan to use INMARSAT capacity to offer global Skyphone service, providing telecommunications for passengers and airline operations. British Airways has begun implementing commercial aeronautic telephone service, which should be fully operational in 1989. The *Société Internationale de Télécommunications Aéronautique* (SITA), in collaboration with OTC Australia, Teleglobe Canada, and France's *Direction Générale des Télécommunications* (DGT), plans to develop and provide aeronautical telecommunications service starting in 1989. INMARSAT and "other recognized operators" will be accessed for satellite data and voice communications for aircraft safety needs, along with in-flight passenger telephone service. SITA has been running in-flight data trials since 1987. Prodot, the European Space Agency's (ESA) project for a pan-European land mobile data service, has led a European effort to use INMARSAT capacity for the provision of air traffic control communications. As many as five aircraft of the Ontario Air Ambulance service will use INMARSAT services, after a year-long experimental period which was scheduled to end in May 1989.[55]

INMARSAT has three INMARSAT-2 satellites on order and options on up to six more from a consortium headed by British Aerospace. Operating in the aeronautical mobile satellite R-band, each INMARSAT-2 will serve at least 125 aeronautical channels and use 3 MHz of bandwidth. The first INMARSAT-2s were slated for late-1989 launches on a McDonnell-Douglas Delta II vehicle.[56] Argentina, Kuwait, Singapore, the U.K., the U.S., and the USSR have all agreed to establish INMARSAT aeronautical earth stations.

In October of 1987, the INMARSAT Assembly of Parties signaled its approval to add amendments to the organization's charter to include land mobile services in the INMARSAT service repertoire. INMARSAT's ambitions to deliver land mobile services were encouraged when the 1987 Mobile World Administrative Radio Conference allocated land mobile frequencies in the L band, where, at present, INMARSAT operates alone.[57]

INMARSAT expects that all of its mobile services will grow with the introduction of smaller and less expensive earth stations. The current Standard A terminals for ships cost less than $35,000, one-half of their 1982 price. The 1990s will see the introduction of the space-segment power efficient, all-digital Standard B earth station. The portable System C, small enough to fit in a shopping bag, began text-only burst transmission at 600 bps in 1989. Following the System C specifications, INMARSAT has created the Enhanced Group Call system which will transmit messages to user-specified terminals, whether on ship or on land. At $5,000, the System C's low price and its capabilities should open up new markets for light vessels and land-based users.

Indicative of the importance which INMARSAT assigns to all mobile satellite services--on land, in the air, and at sea--is the West German idea that the system change its name to the International Mobile Satellite Organization, or INMOBSAT.[58]

Regional Systems

In addition to the global systems, three satellite systems serve the needs of organizations or countries on a regional basis. Two full-fledged and one lesser regional system now compliment INTELSAT: EUTELSAT, ARABSAT, and the Indonesian Palapa satellite system.

EUTELSAT

In 1982 EUTELSAT became the first operational regional satellite communication system. EUTELSAT's Definitive Agreement came into force in 1985, eight years after seventeen members of the European Conference of Posts and Telecommunications (CEPT) signed the Interim EUTELSAT Constitution Agreement.[59]

In accordance with the terms of Article XIV of the INTELSAT Convention, the signatories first notified INTELSAT of their intention to create EUTELSAT. There ensued serious discussions within INTELSAT as to whether the proposed system would cause significant economic harm to INTELSAT. However, the argument that Europe had access to sufficient terrestrial capacity to duplicate the same applications that were being proposed and that INTELSAT did not at the time carry intra-European traffic won the day.

The founding member states of EUTELSAT were Austria, Belgium, Cyprus, Denmark, Finland, France, Germany (FR), Greece, Iceland, Ireland, Italy, Liechtenstein, Luxembourg, Malta, Monaco, Netherlands, Portugal, San Marino, Spain, Sweden, Switzerland, Turkey, United Kingdom, Vatican City, and Yugoslavia. When EUTELSAT's Operating Agreement came into force, the four largest shares in the system were held by France (16.40%), United Kingdom (16.40%), Italy (11.48%), and West Germany (10.82%).[60] The next highest share was that of the Netherlands, with 5.45%.[61]

EUTELSAT has a structure similar to INMARSAT, with an Assembly of Parties, a Council, and a General Secretariat. The Assembly, to which all member states belong and in which each state has a single vote, has responsibility for general policy, determines principles concerning procurement, and is responsible for relations with other organizations. The EUTELSAT Council, with a membership based on investment in the system, is responsible for managing the system and setting rates for services. EUTELSAT's General Secretariat is internationalized and located in Paris.[62]

Each PTT sets its own usage rates. There are, accordingly, wide divergences in rates charged to European program providers on the system's main television satellite, F1.[63]

EUTELSAT provides voice, data, and television services to 26 European member nations. The organization's satellite multi-service system (SMS) for businesses offers high-speed data transmissions, audio and videoconferencing, remote printing of newspapers, facsimile, and other services. The system presently transmits twelve television programs to eighteen countries. EUTELSAT is studying a unique European direct broadcast system, along with mobile communications techniques and services.

Two Ku band satellites serve the EUTELSAT membership, the EUTELSAT I-F1, launched in 1983, and the EUTELSAT F-2, launched in 1984. The I-F1 is used almost exclusively for the distribution of multi-language programming to cable TV systems throughout western Europe. The F-2 is used primarily for voice and data transmission. EUTELSAT also leases space on the French Telecom I. About 180 non-standardized Earth stations operate as receive-only stations for EUTELSAT TV signals, and in mid-1986 there were eighteen standard earth stations.[64]

In March, 1986, EUTELSAT also inaugurated a fully digital satellite telephone network. And in April, 1986, EUTELSAT signed a $225 million contract with *Aerospatiale* to build three EUTELSAT II satellites to be used for business services, telephone, and the distribution of television programs.[65]

EUTELSAT may take up responsibility for Prodot, the European Space Agency's project for a pan-European land mobile data service. Although it is not clear whether EUTELSAT now has the capacity to handle the service's potential users, primarily 400,000 to 800,000 trucks, PTT policies will exercise great influence over Prodot's future. Speculation is that "some of the PTTs would be jealous of land mobile service" and see it as a threat to their own systems. The same

European source adds, however, that the PTTs, as owners of EUTELSAT, should realize that they would profit from any of its success. Additionally, the European climate seems to favor a unified land mobile service, be it terrestrial or satellite. To date thirteen European countries have agreed to establish a single digital cellular system. The European Green Paper on telecommunications calls for improved mobile communications, and an old PTT bias against satellites has apparently vanished.[66]

ARABSAT

ARABSAT was first discussed by the League of Arab States, and the decision to go ahead with the regional system was made in 1980. ARABSAT supported the case for its compliance with the requirements Article XIV of the INTELSAT Convention by arguing that ARABSAT would carry entirely new traffic and would therefore not cause severe economic harm to INTELSAT. Furthermore, ARABSAT stated that if INTELSAT refused its request, it would construct a terrestrial network that would accomplish the same task.

The charter members of the system were Algeria, Bahrain, Djibouti, Egypt, Iraq, Jordan, Kuwait, Libya, Mauritania, Morocco, Oman, the PLO, Qatar, Saudi Arabia, Somalia, Sudan, Syria, Tunisia, Yemen Arab Republic, Yemen (People's Democratic Republic) and the United Arab Emirates. Reported 1985 ownership figures showed that Saudi Arabia held a 50% share, followed by Kuwait with 18% and Libya with 8%.[67]

ARABSAT immediately ran into difficulties. Several transponders on ARABSAT FI failed, and the satellite had difficulties in station-keeping. Members were also slow to pay their contributions and to build ground stations to access the system, probably because of the loss of oil revenue that occurred subsequent to the decision to go ahead with the system. In addition, when Egypt was excluded from the League of Arab States in 1979, ARABSAT lost a major source of funds and a major contributor of programs. Finally, ARABSAT's first General Manager was dismissed for cause.

Whether ARABSAT lives up to the hopes of its members will depend on a number of factors, not the least of which is the future price of Middle East oil. Other factors include whether or not ARABSAT can lower its transponder cost sufficiently to attract members that currently belong to INTELSAT.

ARABSAT at least benefits from the pride which the League of Arab States takes in its operation. Of special significance was the decision of Oman to move its leased transponder on an Indian Ocean

INTELSAT satellite to ARABSAT2 in late 1986 and the 1989 decision to reinstate Egypt as a member of the Arab League.[68]

The Palapa System

Palapa, the third regional system, is of significance not only for what it is but as a possible model for other satellite communication systems which originate as national systems. Palapa was originally proposed by the president of Indonesia in 1972 as a way to improve the standard of living across the country's 13,677 islands. The system became operational in 1976 and began leasing capacity to the Philippines in 1978. Thereafter, Thailand and Malaysia began to use the system, and in 1979 INTELSAT formally permitted the system to operate on a regional scale. The Indonesian government remains the system's sole owner.[69]

The system's first satellite, Palapa A1, was launched in 1976 and is no longer in service. In 1983, the first of Indonesia's second generation satellites, the Palapa B1 C band satellite, was successfully deployed. This satellite provides a variety of voice, video, and high speed data services to the islands of Indonesia and its neighbors.

Expanding Systems and New Regionals

Expansions are underway in the EUTELSAT, ARABSAT and Palapa regional systems, and several other regional systems are in the planning stage, including the Scandinavian Nordsat project, Africa's Afrosat system, and an Andean system for Brazil, Chile, Colombia, Ecuador, and Peru.[70]

NATIONAL SYSTEMS

Although most developed countries are served by INTELSAT, and some by regional systems as well, several have also implemented their own domestic systems. As entities limited to the provision of domestic telecommunications services, national systems cannot be termed "competitors" to INTELSAT, for INTELSAT's founding mission is the provision of international telecommunications. However, there is a good possibility that these domestic systems may expand their services to other countries in their region as has been done with Indonesia's Palapa system.

With the modernization and digitalization of public telecommunications networks taking place around the world, it is significant to note that many countries have plans for the increased use of communications satellites in their future networks. In addition to Indonesia, the U.S. and the U.S.S.R., countries with their own satellite systems or systems in the planning stage include Japan, Canada, Australia, France, West Germany, Italy, China, India, Mexico, Brazil, and the U.K.[71] A discussion of some of these systems follows.

Japan

The 1983 launches of the CS-2A and CS-2B satellites introduced commercial satellite communications into Japan's domestic network. The CS-2s have their roots in the experimental CS satellite, launched in 1977. The CS-2 system operates in the C and Ka bands. The CS-2A was the first commercial satellite to operate in the Ka band. The CS-2B functions as an in-orbit spare and can provide new Japanese digital communications services. The Nippon Telegraph and Telephone Corporation (NTT) uses four of the CS-2A's Ka band transponders and both of its C band transponders. The CS-2A provides redundancy and peak-time capacity to Japan's terrestrial network, as well as providing service to remote areas. Mobile earth stations can broadcast live television signals to network studios via the CS-2A.[72]

The main objective of the CS-2 system is to develop and utilize satellite technology. The Japanese government provides 40% of the CS-2 budget while users provide the remaining 60%.[73]

The CS-3, the successor to the CS-2, is under development. Two CS-3 satellites are scheduled for launch in 1992 and 1994. Each is to carry one C band and ten Ka band transponders, with a total capacity of 6,000 voice channels.[74]

Mitsubishi Electric Corp., Mitsubishi Trading Corp., and Milco have joined together to construct, deploy, and operate a private domestic satellite system for Japan. The two planned SpaceSats will follow the design of the INTELSAT V and V-A spacecraft bus, originally manufactured by Ford Aerospace for INTELSAT. Ariane 4 vehicles are being used to launch the satellites. One of the SpaceSats will function as an in-orbit spare, and each will hold 12 Ku band and 23 Ka band transponders. Thirteen of the Ku band transponders will deliver a powerful spot beam over a 124 mile diameter region.[75]

Along with the Canadian Hermes, the Japanese Broadcast Satellite Experiment (BSE) pioneered DBS services in the Ku band.[76] Launched in 1978, the BSE satellite performed three years of experiments with high-definition television and small-aperture antennas from 1 to 1.6 meter diameters. All three of the satellite's high-power transponders failed prematurely, however. The Japanese government decided to continue the DBS effort with the BS-2 series of satellites designed to transmit to 400,000 subscribers whose isolated geographical locations render terrestrial means impractical. The satellites carry the programming of Japan's national television network, *Nippon Hoso Kyokai* (NHK), which provides the uplink from its Tokyo studios. The BS-2A and the BS-2B satellites, deployed in 1984 and 1986, became the first satellites to fully adhere to the DBS specifications reached at the 1977 World Administrative Radio Conference of the ITU. The BS-2B contained several improvements over the BS-2A, which suffered the failure of two of its three transponders only three months into operation.

The two BS satellites serve the islands of Japan and broadcast an 11 GHz signal of at least 45 dBW. On the main island, the signal can be received by antennas with diameters as small as 70 centimeters. Elsewhere in Japan, 1.8-meter antennas are necessary, while 3-meter antennas assure reception in South Korea.

Ninety-percent of the system's broadcasts duplicate those of the Japanese terrestrial network, although NHK opened discussions with the government in 1986 to make major changes in the programming on the second of the BS-2B's three transponders. There is minimal interest in expanding DBS service to non-remote areas.[77]

Canada

Canada's role in satellite communications dates to the 1962 launch of Alouette 1, a scientific ionospheric satellite put into orbit by NASA. The first domestic satellite communications company, Telesat Canada, was incorporated by the Canadian parliament in 1969. In 1972, Canada placed Anik, the world's first geosynchronous domestic satellite, into geostationary orbit. The Anik B1 satellite was also the first to employ dual-band technology, operating in both the C and Ku bands. Canada, with the help of U.S. agencies, has launched fourteen satellites without a major launch or operational malfunction.[78]

With its vast geographic area and low population density, Canada has been well served by the technology of communication satellites; many remote regions are now receiving telephone and television service that would be difficult or too expensive without the use of satellites. In the more populated southern regions, satellites are also used for TV program distribution, as is the case in the U.S. Educational TV, radio, and multilingual services are also available.[79]

Telesat Canada plans to continue communications satellite research and development, employing many of the new technologies being used in other countries. Of particular interest is the mobile satellite service (MSAT) planned for the mid-1990s.[80]

Australia

The Aussat A1 and A2 satellites were launched in 1985 to provide a variety of communications services throughout Australia, including DBS programming to homes in remote outback locations, regional radio broadcasts, multicultural and educational television programs, new voice and digital data services, teleconferencing, domestic airline communications between controllers and from ground to air, and television programming for hotels and clubs. The Homestead and Community Broadcasting and Satellite Service is extending the reach of the state-owned Australian Broadcasting Corporation's programming to 350,000 residents in remote rural areas. The service has also improved the television reception of a million additional Australians. The Remote Community Television Service provides commercial programming to regions in the Outback. Individuals in remote regions can receive broadcasts with antennas 1 to 1.5 meters in diameter, or join with others using community-owned satellite antennas 2.4 to 3 meters in diameter. Coaxial cable or low-power TV transmitters then bring the television signal to members of the community.

Hughes Aircraft manufactured the Aussat satellites. They operate in the Ku band and are backed up by an in-orbit spare, the Aussat A3, deployed in 1987. Two beams in the Aussat 3 are designed to provide telecommunications services to New Zealand and other nations in the southwestern Pacific. Each Aussat satellite holds 15 transponders and carries three pairs of antennas, providing nationwide and spot coverage. An on-board microwave switching matrix controlled from the ground allows transponders to connect to any beam, thereby optimizing the allocation of satellite resources.[81]

Aussat Proprietary Limited, a wholly government-owned satellite operating company, manages the system. Aussat's Memorandum and Articles of Association and the Australian Satellite Communications Act of 1984 provide the framework within which Aussat will operate the satellite system. The Act makes it clear that satellite services are to be provided as an integrated part of Australia's national telecommunications infrastructure.

France

France's first domestic satellites, the Telecom series, commenced service with the 1984 launch of Telecom F1. Telecom F2, also known as Telecom 1B, followed in 1985, and Telecom F3 in 1987. All of the Telecom satellites are designed for operational lives of seven years. The satellites had originally been planned for 10° and 7° west, and 5° east longitude positions, but coordination with INTELSAT and

INTERSPUTNIK satellites resulted in shifting all three satellites 2° to the east. Potential interference problems also led INTELSAT to request that the French limit C band television broadcasts to the first of each satellite's twelve transponders. In addition, the French complied with INTELSAT's request to restrict Telecom F2's broadcasts to the Ku and X bands (8-12 GHz), foregoing C band transmissions which could have interfered with INTELSAT operations.[82]

The French aerospace firm *Matra* was the primary Telecom satellite contractor, and subcontractors included *Thompson CSF*, Ford Aerospace, and SNIAS. The Telecom system has assumed many of the services handled by the successful Symphonie satellites, Europe's first communications satellites launched in the mid-1970s. The Symphonie satellites were among the first three-axis geosynchronous spacecraft. They eventually exhausted the fuel that enabled the thrust to keep them in-line with their stations, forcing ground stations to keep track of figure-eight orbits over the equator. Satellite tracking capabilities are not provided by the ground stations utilized for the Telecom satellites, which have consequently been designed to maintain geostationary orbits within plus or minus 0.1 degree of their assigned positions over the equator.[83]

Both satellites feature four C band transponders, three of which connect to semi-global beams whose end-of-life signal strength varies from 26.5 to 29.5 dBW. These beams create a footprint over French territories in the Caribbean and in the Americas, and the C band transponders on the Telecom F1 carry most of the system's telecommunications traffic. Telecom F2 provides radio and television programming both within France and abroad. The other C band transponder on each satellite uses a spot beam to effect service between the French telephone network in Europe and private branch exchanges (PBXs) in South America and the Caribbean. Communications services to diverse locations within the European Economic Community (EEC) can be routed through six Telecom transponders using spot beams in the Ku band. The Ku band transponders on the Telecom F1 are connected to a spot beam which covers France, whereas those on Telecom F2 connect to six spot beams directed at France.[84]

The French TDF-1A and TDF-1B satellites, along with their virtual double, the German TV-Sat, are now initiating European DBS services. European broadcasting organizations will be able to access DBS capacity through the French government's DBS marketing company, the *Société de Commercialisation des Satellites de Télédiffusion*.

The DBS satellites of both countries will operate for 7.5 years and are made by *Eurosatellite*, a common subsidiary of *Aerospatiale* and *Thompson-CSF* in France, Germany's *Messerschmitt-Bolkow-Bohmm* (MBB) and *AEG-Telefunken*, and ETCA of Belgium. The German and French versions will incorporate different transponder frequency schemes, polarities, and antenna designs, but both will feature common subsystems and modular components. High power 230 or 260-watt traveling wave tube amplifiers will enable small satellite dishes two feet in diameter to receive noise-free, 63 dBW signals from the Eurosatellite birds. Both TDF and the TV Sat utilize the D2-Mac (Multiplex Analog Component-Type D2) video standard, the baseband of which is sufficiently narrow to be transmitted on existing cable networks in Europe. The TDF-1 satellite enables the ground operating center to control the position of the on-board dish within a pointing accuracy of 0.1 degree.

Germany

The *Deutsche Bundespost* uses transmission capacity on both the French Telecom 1 and European ECS-F2 satellites to transmit data at up to 2 Mbps, including computer-to-computer traffic, whole-page newspaper transmission, decentralized printing, video teleconferencing, and speech-accompanied text and data traffic. In addition, the Bundespost leases INTELSAT television transmission service for news, sports, and variety programming of the West German 3-Sat ZDF, Swiss SRG, and Austrian ORF networks.[85]

The German PTT has launched its *Deutscher Fernmelde Satellit* (DFS) 1 and 2 satellites, and has plans for a subsequent, optional DFS satellite. One of the satellites functions as an in-orbit spare, and the DFS system will terminate at thirty-four ground stations, providing high-speed digital transmissions of teleconferencing, telephony, voice, facsimile, and data information, along with cable-TV services and the interconnection of network video studios across Germany.[86]

The DFS satellites are designed for ten-year lives and carry eleven transponders each, ten of which will operate in the Ku band and the remaining transponder utilizing the Ka band. Five of the Ku band transponders will connect five regional television services to local cable network distribution points. Digital business services will be carried on two other Ku bands linked to 3.5 to 4.5-meter diameter satellite dishes loaded with 250 W transmission amplifiers. The business dishes will incorporate switching equipment capable of providing access to 100 users connected by subscriber lines. The system will employ two reference earth stations to ensure proper synchronization, acquisition, and demand assignment in the system's TDMA modulation scheme. Two Ku band transponders and one Ka band transponder will deliver point-to-point communications between large diameter earth stations. These transponders will be 90 MHz wideband, each with capacity sufficient for two high-quality TV channels or one duplex 140 Mbps digital communications link. The latter will be able to handle up to 1,920 simplex channels, reduced to 1,440 in a forward error correction mode. The minimum signal level for domestic reception will be 48 dBW.

The German PTT will use the Ka band capacity to test the use of VSATs to uplink live events. The Ka band capacity will find subsequent assignment to commercial traffic.[87]

As mentioned in the summary of French satellite programs, West Germany began DBS services with its 1989 launch of TV-Sat following the 1987 failure. The TV-Sat carries up to five television channels. Government broadcasters operate two of the channels, and private broadcasters are assigned the rest.[88]

Other Systems

United Kingdom

As a leading signatory to the INTELSAT, EUTELSAT and INMARSAT organizations, British Telecom has a vested interest in ensuring that these organizations have the facilities in place to satisfy the needs of users in a cost effective manner.

The INTELSAT and EUTELSAT systems, plus Telecom 1, are expected to provide coverage of the U.K., with extensions to Scotland, Ireland, and offshore structures in the North Sea. Services are expected to include TV distribution as well as voice and data transmission.[89]

Italy

Italy's ITALSAT program is being undertaken with the intention of designing and constructing an experimental domestic telecommunication satellite.

ITALSAT is designed to experiment with digital domestic links at 30/20 GHz, either via a multibeam regenerative payload or via a transparent country-wide coverage payload. A third payload will allow propagation experiments at 40 and 50 GHz.

The main telecommunication network architecture is based on six spots coverage of the Italian territory and a few earth stations in each spot. The network will experiment with digital voice at 32 kbps and data at up to 8 Mbps in point-to-point and point-to-multipoint applications.

A second, less advanced telecommunications network will be available on a nationwide coverage basis for business communications services. Capabilities demonstrated by the ITALSAT system could contribute significantly to a more rapid transformation of the national telecommunication network towards the ISDN configuration.[90]

China

The People's Republic of China has two experimental satellites in geostationary orbit, the STW-1 and STW-2. The first satellite was launched in 1984, the second in 1986. Both satellites operate at C band, and are used to conduct various television and radio transmission experiments. China plans to follow up these experimental satellites with a commercial communications satellite in the 1990s.[91]

India

The Indian National Satellite (INSAT) program is a joint venture among several departments of the Indian government. Two satellites of the Insat system have been launched to date. The first satellite, placed into orbit in 1982, encountered several technical problems and is no longer providing service. The second satellite, launched in 1983, has made a significant contribution to the telecommunications capabilities of the Indian subcontinent. Insat IB operates in U, S, and C bands and is providing voice, data and video services for a variety of uses throughout India.[92]

Mexico

The Morelos F1 and F2 (an in-orbit spare) satellites, launched in 1985, provide television distribution services throughout Mexico. In addition, this satellite system will be utilized to expand Mexico's rural telephony network, and will also be used for data transmission services. Both satellites are capable of transmitting in either the C or Ku band.[93]

Brazil

With the assistance of the Canadian government, Embratel, the government-owned telecommunications entity of Brazil, contracted with Hughes Aerospace and SPAR for the design of two customized satellites. These satellites were deployed in 1986, and comprise the beginnings of the Brazilsat system. The satellites are used for radio, television and voice transmission throughout Brazil and several neighboring countries. The Brazilsat system operates in the C band.[94]

SUMMARY CONCLUSIONS

The United States plan of the 1960s was to create a single global system providing satellite communications for the world at large. The American plan, however, was doomed from the start. It was wishful thinking to believe that all other countries would voluntarily join a system in which the United States held a dominant position. It was also wishful thinking to believe that other countries, individually or in groups, would not someday catch up with the necessary technology and decide to build satellite communication systems of their own.

INTELSAT soon had a rival in INTERSPUTNIK, created by the USSR, which the other members of the Socialist Bloc were quick to join. Then came the maritime nations with INMARSAT, the Europeans with EUTELSAT, the Arab nations with ARABSAT, and Indonesia with its Palapa system. The more industrialized nations, and some of the not so industrialized nations, have also created domestic satellite communication systems, some of which carry interregional traffic.

Nevertheless, U.S. policy had a tremendous impact on the evolution of satellite communications in its earlier years and INTELSAT, while not all that the U.S. wanted in the beginning, became a strong and significant factor in satellite communication in general. Even today little can be done without taking into account the interests of INTELSAT, and there are few indications that the members of INTELSAT would have it any other way.

NOTES

1. For an account of the early years of satellite development see Burton I. Edelson, "The Experimental Years," in Joel Alpter and Joseph N. Pelton (eds.), *The Intelsat Global Satellite System* (New York, NY: American Institute of Aeronautics and Astronautics, Inc., 1984), pp. 39-53.

2. See *Statement by the President on Communication Satellite Policy*, July 24, 1961.

3. For a detailed account of the negotiations that preceded the passing of the Communications Satellite Act of 1962 see Jonathan F. Galloway, *The Politics and Technology of Satellite Communications* (Lexington, MA: D.C. Heath and Company, 1972), Chapter 4. See also Patricia Jefferson and Johan Benson, "Edward C. Welsh: A Personal History of the Comsat Act," *Satellite Communications*, Jan. 1979, pp. 22-25.

4. See U.S., *Satellite Communications Act of 1962*, Sections 303 & 304. The U.S. international carriers soon divested themselves of Comsat stock and their twelve members of the Board were henceforth elected by the general public shareholders.

66

5. *Ibid.*, Sec. 201.

6. *Ibid.*, Article 12 (C). In 1979 Comsat handed essentially all technical facilities over to the INTELSAT organization.

7. INTELSAT, *Agreement Relating to the International Telecommunications Satellite Organization, "INTELSAT,"* Washington, D.C, August 20, 1971, Article XIV (d). This paragraph reads: "To the extent that any Party or Signatory or person within the jurisdiction of a Party intends individually or jointly to establish, acquire or utilize space segment facilities separate from the Intelsat space segment facilities to meet its international public telecommunications services requirements, such Party or Signatory, prior to the establishment, acquisition or utilization of such facilities, shall furnish all relevant information to and shall consult with the Assembly of Parties, through the Board of Governors, to ensure technical compatibility of such facilities and their operation with the use of the radio frequency spectrum and orbital space by the existing or planned Intelsat space segment and to avoid significant economic harm to the global system of Intelsat. Upon such consultation, the Assembly of Parties, taking into account the advice of the Board of Governors, shall express, in the form of recommendations, its findings regarding the considerations set out in this paragraph, and further regarding the assurance that the provision or utilization of such facilities shall not prejudice the establishment of direct telecommunication links through the Intelsat space segment among all the participants."

8. For an interesting and informative account of Article XIV(d) and its background, see Irwin B. Schwartz, "Pirates or Pioneers in Orbit? Private International Communications Satellite Systems and Article XIV(d) of the Intelsat Agreements," *Boston College International and Comparative Law Review*, Vol. IX, No. 1 (Winter 1986), pp. 199-242.

9. The U.S. argued that INTELSAT should be restricted to ordinary communications services such as telephone, telegraph, telex, facsimile, data, and point-to-point broadcasting.

10. *Agreement Relating to the International Telecommunications Satellite Organization, "INTELSAT"*, Arts. VIII & XVII.

11. *Ibid.*, Art. VII.11.

12. *Ibid.*, Art. IX.

13. See Henry M. Rivera, "Separate Systems and International Communications," *IEEE Communications Magazine*, Vol. 25, No. 1 (Jan. 1987), p. 39.

14. Dean Burch, "INTELSAT: The Tomorrow Organization," in Pelton and Howkins, p. 25.

15. INTELSAT, *Annual Report, 1979*, Washington, D.C., 1979, pp. 3 & 24 and *INTELSAT Annual Report, 1987-88*, pp. i, 2, 9, 19, 20, 39, & 40.

16. INTELSAT, "Intelsat to Provide Global Business Communications Service," Sept. 21, 1983; and INTELSAT, Board of Governors, *Annual Report by the Board of Governors to the Eighteenth Meeting of Signatories on Intelsat Services*, Washington, D.C., 16 April, 1988, pp. 6-8.

17. *Ibid.*, p. 7.

18. *Ibid.*

19. *Ibid.*, pp. 10-11, Attachment No. 5, pp. 1-2; and INTELSAT *Annual Report, 1987-88*, p. 23.

20. Burch, p. 25.

21. INTELSAT, *Annual Report by the Board of Governors*, (1988), p. 9.

22. *Ibid*, p. 5. This is important in that without satellite backup submarine cables would have had about a 93% availability service rating.

23. *Ibid.*, p. 27.

24. *INTELSAT Annual Report*, 1986-87, p. 32.

25. INTELSAT, *Annual Report of the Board of Governors* (1988), pp. 3-4.
26. *Ibid.*, pp. 16-17.
27. *Ibid.*, p. 17.
28. *Ibid.*, pp. 12-13.
29. *Ibid.*, p. 13.
30. *Ibid.*, p. 14.
31. *Ibid.*
32. *Ibid.*, p. 16.
33. *Ibid.*, pp. 15-16.
34. Peter K. Runge and Patrich R. Trishitts, "Future Underseas Light Wave Communications Systems," *Signal Magazine*, June 1983, p. 31. With digital circuit multiplication this capacity may expand to 50,000 circuits.
35. Pelton and Howkins, p. 128.
36. *Ibid.*
37. Long, p. 83.
38. Theo Pirard, "INTERSPUTNIK: The Eastern 'Brother' of Intelsat," *Satellite Communications*, Aug. 1982, p. 39.
39. INTERSPUTNIK's traffic is only about 0.5% of that of Intelsat.
40. See Chapter Three for a further discussion of the possible effects that competition may have on the Intelsat system.
41. John Carver Scott, "Back in the USSR," *Satellite Communications*, April 1988, p. 44.
42. Pelton and Howkins, pp. 129-130.
43. Howell, p. 252.
44. INMARSAT, *Annual Review 1987/1988*, p. 4.
45. Josentuliyan, "The Establishment of an International Maritime Satellite System," *Annals of Air and Space Law*, Vol. 2, 1977, p. 327.
46. Pelton and Howkins, p. 130.
47. Olof Lundberg, "INMARSAT on the Move," in Pelton and Howkins, p. 29.
48. See Long, p. 108 and Pelton and Howkins, p. 131.
49. The coast earth stations will also be capable of handling aeronautical communications.
50. Pelton and Howkins, p. 132.
51. Theo Pirard, "New Markets for the New INMARSAT," *Satellite Communications*, March 1988, p. 21.
52. *Ibid.*, p. 20.
53. INMARSAT, *Annual Review*, 1987-88, p. 12.
54. *Ibid.*, p. 10.
55. *Ibid.*, p. 12.
56. See Long, pp. 108-110 and Pirard, pp. 20-22.
57. Tamara Bennett, "Prodat Enters Europe's Mobile Fray," *Satellite Communications*, March 1988, p. 26.
58. Pirard, "New Markets . . . ," p. 21
59. Theo Pirard, "EUTELSAT: Satellites Linking Europe," *Satellite Communications*, July 1983, p. 35.
60. Simone Contreix, "EUTELSAT: Europe's Satellite Telecommunications," *Regulation of Transnational Communication*. Michigan Yearbook of International Studies (New York: Clark Boardman Company, Ltd., 1984), p. 95.
61. Pelton and Howkins, p. 136.
62. Pirard, "EUTELSAT... ," p. 35.
63. Howkins and Pelton, p. 136.
64. Pelton and Howkins, pp. 136-137.

68

65. Theo Pirard, "Competition in Europe?," *Satellite Communications*, Aug. 1986, p. 22.

66. Bennett, p. 26.

67. Pelton and Howkins, p. 135.

68. On the problems confronting ARABSAT, see Lofti Maherzi, "A Highway Into Space," *Intermedia*, Vol. 14, No. 6 (Nov. 1986), pp. 19-21 and Long, pp. 123-125.

69. Pelton and Howkins, p. 192.

70. See Howell, p. 255.

71. *NTIA Telecom 2000*, p. 274.

72. Long, pp. 480-482.

73. Shinji Matsumoto, "Present and Future Domestic Satellite Communication Systems in Japan," *Conference Proceedings*, *Fibersat 86*, pp. 43-49.

74. Long, p. 482.

75. *Ibid.*, pp. 486-487.

76. *Ibid.*, p. 6.

77. *Ibid.*, pp. 492-495.

78. Barry Murphy, "Satellites in Canada: Past, Present, and Future," *Conference Proceedings, Fibersat 86*, p. 82.

79. *Ibid.*, p. 83.

80. *Ibid.*, p. 84.

81. Long, pp. 465-492.

82. *Ibid.*, pp. 173, 177-178, 167-168 & 180.

83. *Ibid.*, pp. 176-177.

84. *Ibid.*, pp. 173-180.

85. OECD, *Trends of Change in Telecommunications Policy*, (Paris: OECD, 1987), p. 248; and Long, p. 120.

86. Long, p. 134.

87. Long, p. 134-136.

88. Long, p. 204.

89. OECD, pp. 330-331.

90. *Ibid.*, p. 269.

91. Long, pp. 489-491.

92. *Ibid.*, pp. 509-515.

93. *Ibid.*, pp. 387-390.

94. *Ibid.*, pp. 262-265.

CHAPTER THREE

U.S. POLICY OF GLOBAL COMPETITION

INTRODUCTION

It is difficult to determine exactly when the U.S. became disillusioned with its policy of global monopoly in satellite communications and its creation, INTELSAT, and began to favor the entry of private U.S. satellite communications providers. Certainly the United States was unhappy with the feeling of many INTELSAT members that, as long as COMSAT remained the chief manager of the system and its largest shareholder, U.S. control over INTELSAT was excessive. This did not, however, prohibit the parties to the Interim Agreement from arriving at a new agreement in 1971 which perpetuated COMSAT's role as a management service contractor for an additional six years. Neither did it prohibit COMSAT from remaining the major technical contractor to INTELSAT for a number of additional years after that.

The first overt manifestation of a change in U.S. telecommunications satellite policy was the decision on the part of the Nixon administration to permit private concerns to create their own domestic satellite communications systems. The second, and more important, development was the decision of the Reagan administration to authorize private companies to create satellite systems to compete in the provision of long-distance international communications. Other policy shifts were soon to follow.

Changes in U.S. policy towards INTELSAT in the 1970s and 1980s did not so much reflect U.S. dissatisfaction with INTELSAT as it did a devaluation of the importance that the U.S. assigned to it. During those decades the U.S traffic load on the system dropped from approximately 60% to less than 25%, and INTELSAT became less reliant on U.S. technology and research. With the implementation of the Permanent Agreement and the creation of the Assembly of Parties and the Meeting of Signatories, the U.S. had to adapt to the increased political influence of members with small shares in INTELSAT. The reduced U.S. interest in INTELSAT was also due to a domestic political development, the trend toward deregulation which began under Carter in the civil aviation and trucking industries and continued into the Reagan years with a focus on both domestic and international telecommunications.[1]

While it is too early to predict what will be the final outcome of this changing U.S. policy, it is important at this juncture to take account of the fact that one of the more important actors has made a basic change in its policy, a change likely to have a tremendous impact on the future development of international satellite communications.

This chapter will investigate the reasons for the change in U.S. policy from one of global monopoly to one of global competition, the reactions of the other major players to this change, and some of the effects that the policy shift might have on satellite communications in the future.

THE "OPEN SKIES" DECISION

The Communications Satellite Act and Private Systems

There is little doubt that the framers of the Communications Satellite Act of 1962 intended that the planned system be a world-wide monopoly to be used principally, perhaps even exclusively, for international communication. This emphasis was contained in speeches by both Presidents Eisenhower and Kennedy.[2] Further, as stated earlier, the leading paragraph of the Communications Satellite Act of 1962, Declaration of Policy and Purpose, states categorically "that it is the policy of the United States to establish, in conjunction and in cooperation with other countries, as expeditiously as practicable a commercial communications satellite system, as part of an improved global communications network . . . ."[3]

The technology appeared ideal for long distance communication and, besides, many felt that the United States already had the best possible domestic communication system in the form of the largest, most efficient government regulated domestic monopoly, AT&T. If the U.S. was satisfied with its terrestrial communication system, they argued, shouldn't other countries feel the same?

There was, however, enough flexibility in the Act to permit a different interpretation. The fourth paragraph of the Act's Declaration of Policy and Purpose states:[4]

It is not the intent of Congress by this Act to preclude the use of the communications satellite system for domestic communication services where consistent with the provisions of this Act nor to preclude the creation of additional communications satellite systems, if required to meet unique governmental needs or if otherwise required in the national interest.

One of the first attempts to use that flexibility to create a domestic satellite communications system was that by the American Broadcasting Company. In collaboration with Hughes Aircraft, ABC applied to the FCC on September 21, 1965, for permission to create a satellite system to distribute network television programs to affiliates throughout the United States. The purpose, of course, was to allow ABC to avoid AT&T's high-cost, long-distance rates.

After COMSAT filed in opposition, the FCC returned the ABC request without prejudice, including the notation that:[5]

Your application proposed a use of space techniques which is outside the purview of the established rules of the Commission. Furthermore, the unique nature of the proposal presents basic questions of law and policy which must be resolved before a proposal such as yours could be considered.

What followed was a prolonged struggle involving the FCC, Congress, the White House, and the communications industry. Proponents of domestic satellite systems argued, in general, that the granting of such licenses would be consistent with the U.S. philosophy of free enterprise and that free competition would lower rates for businesses. Opponents argued that the 1962 Act granted COMSAT the exclusive right to offer international and domestic satellite services and that the existing terrestrial system was adequate to the task.

An underlying issue in all of the debates, however, was the role that was to be played by AT&T. While Bell's position as a government-regulated monopoly seemed to be fairly secure in the 1960s, there has already been indications that it would not always remain so. The first cracks in the structure appeared in the late 1950s with the Hushaphone and the Above 890 microwave decisions. The former permitted attachments to the telephone for the first time and the latter allowed large businesses to use microwave transmission for private lines. During the debate on domestic satellite communication systems, a further deterioration of AT&T's position took place with the Carterphone interconnect decision in 1968 and the decision to allow MCI to offer private line services in 1969.[6]

The Rostow Report of 1968/1969, commissioned by President Johnson, was a milestone in the quest for private, domestic satellite communications. In its broad overview of U.S. telecommunications, the Rostow Report devoted a significant section to the problem of domestic satellites. While the writers of the report felt that satellites would be useful in many domestic applications, they also felt that it would be premature to give them a firm blessing at that time. Instead, it was suggested that a pilot program be carried out under the auspices of COMSAT in order to obtain additional data.[7]

Developments During the Nixon Administration

The advent of the Nixon administration brought forth new faces and new priorities. Almost from the beginning, the Nixon Administration began urging the FCC to decide the issue and decide it favorably. In a Memorandum to the Chairman of the FCC dated January 23, 1970, a White House aide stated bluntly: "Government policy should encourage and facilitate the development of commercial satellite communication systems to the extent private enterprise finds them economically feasible."[8] Clay Whitehead, Nixon's head of the new Office of Telecommunications Policy, was just as direct: "There are customers waiting for satellite services and prospective suppliers with capital and the will to offer them on a commercial basis. We see no reason for the government to continue keeping these groups apart."[9]

The U.S. domestic "open skies" policy became a reality in 1972. The justification can be found in the FCC's Second Report and Order of that year in which it is stated that "satellite technology has the potential of making significant contributions to the nation's domestic communications structure by providing a better means of serving certain of the existing markets and developing new markets not now being served."[10] In addition, the Commission felt that "multiple entry is most likely to produce a fruitful demonstration of the extent to which the satellite technology may be used to provide existing and new specialized services more economically and efficiently than can be done by terrestrial facilities."[11]

The objectives to be pursued in granting licenses for the use of satellite systems for domestic communication purposes were:[12]

(a) to maximize the opportunities for the early acquisition of technical, operational, and marketing data and experience in the use of this technology as a new communication resource for all types of services;

(b) to afford a reasonable opportunity for multiple entries to demonstrate how any operational and economic characteristics peculiar to the satellite technology can be used to provide existing and new specialized services more economically and efficiently than can be done by terrestrial facilities;

(c) to facilitate the efficient development of this new resource by removing or neutralizing existing institutional restraints or inhibitions; and

(d) to retain leeway and flexibility in our policy making with respect to the use of satellite technology for domestic communications so as to make such adjustments therein as future experience and circumstances may dictate.

AFTER THE "OPEN SKIES" DECISION

A series of FCC decisions in the 1980s has guaranteed that there will be competition among carriers in the domestic satellite arena. These regulations included streamlining the regulation of the non-dominant carriers. In 1981, the FCC reduced rates for leasing a satellite transponder.[13] The FCC intended these reductions to "promote the development of satellite technology and that of technological innovation." In the Transponder Sales ruling, the FCC authorized Hughes Communications, Inc. to sell twenty-four transponders on its Galaxy I Satellite and RCA American Communications to sell five on its Satcom IV satellites. Western Union Telegraph was also granted the right to market eleven transponders on its satellites.[14]

Another significant area of FCC policy making has been that of earth station deregulation. Veronica Ahern, former chief of the International and Satellite Branch of the FCC's Common Carrier Bureau stated that "the elimination of licensing requirements for TV receive-only antennas was one of the most fundamental decisions made during the past decade."[15] In addition, federal tax laws which provide investment tax credits for transponder purchases and deregulation of small earth stations have facilitated research and development of satellite transponders and VSAT technology.

As a result of FCC actions in the 1980s, there has been a steady increase in competition as the government has moved out and commercial industry has taken hold. This has made it attractive for private carriers to invest in satellite communications systems. Thirty satellites operated by seven different systems account for most domestic U.S. communications satellite transmissions. These systems are operated by Alascom, AT&T, COMSAT, Contel ASC, GE Americom, GTE Spacenet, and Hughes Communications.[16]

Before 1984, when demand for satellite positions exceeded the supply, initial licensing to satellite communications companies was quite liberal.[17] More recently, the FCC allocation process was drastically slowed down due to the increased demand, and the FCC started testing a series of new procedures to permit a more rapid allocation of resources. This was facilitated in November, 1988, when the FCC authorized seven companies to build thirteen new domestic satellites and to launch nine of them. The companies involved included: Alascom, Inc.; AT&T; Contel ASC; GE American Communications Inc.; GTE Spacenet Corp.; Hughes Communications Galaxy Inc.; and National Exchange Satellite Inc. At the same time the Commission authorized the replacement of ten existing satellites, "indicating for the first time an acceptance of replacement rights for companies operating fixed satellites."[18] The new assignments and the replacements will bring the number of in-orbit comsats to a total of forty-two.

According to the FCC, "The newly authorized in-orbit satellites will provide domestic satellite capacity through the 1990s and will offer users a wide range of satellite-delivered services, such as video services, data services including teleconferencing, and VSAT network services."[19]

In a related action, in December, 1986, the FCC mandated that the twelve companies applying for mobile voice and data satellite system authorizations form a single consortium. This ruling was an attempt to speed up the Mobile Satellite Services (MSS) proceeding, which had already lasted four years because of lengthy spectrum allocation proceedings and divergent proposals for system configurations and technologies.[20] In 1988 eight of the original MSS applicants agreed to form the American Mobile Satellite Consortium (AMSC), now called the American Mobile Satellite Corporation. In addition to general mobile service, the American Mobile Satellite Corporation intends to provide aeronautical safety services. The corporation's plans, however, do not anticipate an orbiting system until 1992. Faced with INMARSAT's expanding range of services and non-voice competition from Omninet, and to a certain extent Geostar, AMSC represents a vulnerable $730 million system still on the ground.

The eight members of the AMSC are:[21]

1. Hughes Communications Mobile Satellite Services, Inc.
2. MCCA Space Technologies Corp.
3. McCaw Space Technologies, Inc.
4. Mobile Satellite Corp.
5. Satellite Mobile Telephone, Inc.
6. Skylink Corp.

7. Transit Communications, Inc.
8. North American Mobile Satellite Corp.

Los Angeles-based Omninet was scheduled to begin RDSS, two-way data, and alphanumeric messaging mobile services in June, 1988. By foregoing the ownership of a new system and leasing existing capacity on two transponders of a GTE Spacenet GStar I satellite, Omninet has avoided the regulatory delays suffered by the members of the AMSC. Rather than using L band frequencies allocated to mobile satellite services, Omninet plans to use Ku band frequencies on a secondary basis. Omninet will first go after the trucking market, using circular antennas attached to the roofs of truck cabs.

SEPARATE INTERNATIONAL SYSTEMS DECISION

On March 11, 1983, the Orion Satellite Corporation filed an application with the FCC for the authority to construct, launch, and operate a private international communications satellite system independent of INTELSAT. The proposed system anticipated two geosynchronous satellites, one ground spare satellite, and control earth stations whose primary objective would be to link North America and Western Europe with audio, video, and data services. Orion's proposal has been dubbed a "global communications condominium," since the system would by-pass public switched telephone networks and be limited to interconnection among Orion's clients.

Although the Orion filing was unexpected, its timing was certainly apt and calculated. Orion took advantage of three significant features of regulatory reform: the tendency for the political climate to be more receptive to proposals for international deregulation when it follows domestic deregulation; the reluctance of Congress to transform its intentions into completed legislation in favor of letting the Executive Branch take specific international policy initiatives with the consequent responsibilities; and the preference of Congress to support incremental rather than sweeping changes. Orion exploited the regulatory environment by treating its application as a mere continuation of existing policy and trends, involving minimal government responsibility and action.[22]

The Executive Branch response to the Orion application was a protracted turf war between the State Department and the National Telecommunications and Information Agency (NTIA) in the Commerce Department. Both sides supported the concept of separate systems, but they disagreed about the operational and pricing flexibility that INTELSAT should enjoy in competing against such systems. The

NTIA argued that INTELSAT's charter should be modified to authorize it to set prices with more flexibility. The State Department, on the other hand, held that INTELSAT already enjoyed sufficient pricing freedom as a result of the way it assigned different definitions and thus prices to the same services in different parts of the world. The State Department also asserted that INTELSAT wanted more pricing freedom so that it could subsidize cuts in U.S. rates with revenues from other locations in its far-flung system. The NTIA countered by saying that the INTELSAT's pricing flexibility was insufficient to meet the competition of private systems, and that abuses of cross-subsidization could be prevented. The State Department responded by saying that allowing INTELSAT greater pricing flexibility could provoke its members to act against Orion, because they would end up with the burden of subsidizing users in the U.S.[23]

In November, 1984, President Reagan announced a long-awaited decision that separate international satellite systems would be in the U.S. national interest. This announcement was followed in February, 1985 by the lengthy *White Paper* prepared by the Senior Interagency Group on International Communication and Information Policy which set forth the justification for the decision and explained how it could be accomplished without violating the U.S. obligations to INTELSAT. Absent from the paper was any mention, much less a resolution, of the disagreements which caused the State and Commerce departments to take so long in producing it.[24]

Two major reasons were advanced for the decision: it would result in lower communication costs for American businesses, and it would have a positive effect on the balance of U.S. trade. International telecommunication services, it was stated, cost between two and three times that of comparable domestic services. Private systems could lower the cost to close to that of domestic services, creating a positive influence on American business. As stated in the report:[25]

> Reductions in those communications costs imply lower business--and, ultimately, customer -- costs and expansion in business activity. New entrants may also offer large users services more closely tailored to particular corporate needs. Worldwide credit card and electronic funds transfer operations, for example, may be heavily dependent on the availability of efficient, dedicated satellite communications networks. New communications service options and resulting efficiency gains should be reflected ultimately in lower costs to consumers and, in the case of U.S. firms, enhance the attractiveness of their products in international markets.

As concerns the American trade balance, the Report made the following assertions:[26]

New communications satellite offerings should also have an affirmative effect on the U.S. services sector generally, which is of special importance to U.S. overall foreign trade. In recent years, the services sector has become a major source of export receipts in U.S. balance of payments accounts. Included in this diverse sector are enterprises including data processing, engineering, architectural, and construction services, advertising services, management consulting and accounting services, insurance services, and the provision of video programs, all of which are increasingly dependent on the availability of effective and efficient international communications. The market for U.S. programs is particularly important given the rapid development of cable television, commercial television, and other video service in Europe. In 1982, receipts from services exports were $40.4 billion, about one-fifth the amount of U.S. merchandise exports. Over the past decade, growth in U.S. services exports has partially offset losses in merchandise export accounts. Services constitute a key component of U.S. international trade and expanding U.S. communications options should contribute to its growth.

The report also suggested that private satellite communication systems could have beneficial implications for national defense since the defense establishment relies heavily on international communications services. "A key interest of the Defense Department and the national security community is ensuring the effectiveness and survivability of international communications services through redundant routing and maintaining a broad mixture of international communications facilities." Additional international satellite communication facilities, continues the Report, "would contribute to the 'mix of media' national defense requires."[27]

For these reasons, it was necessary to abandon the older concept of a global monopoly for a new one of made up of INTELSAT and private U.S. satellite communications systems. INTELSAT would continue to do what it does well. INTELSAT, according to the White Paper, "serves the world well. It has established and currently operates an efficient global communications system; promotes closer ties among non-communist countries; facilitates international business expansion; helps to develop markets for U.S. industry; prevents the spread of a global communications satellite network controlled by the

Soviet Union; and is an effective international organization reflecting shared technical and political interests."[28]

In order to protect INTELSAT, the new private systems would not be permitted to engage in public-switched message traffic. As pointed out in Chapter Two, the public-switched service comprises the vast majority of INTELSAT service and it is predicted to increase throughout the 1989-2000 period. Barring the new entrants from the public-switched service reduces any likelihood of significant adverse economic impact on INTELSAT.[29] As stated in the White Paper, "New entrants . . . should be limited to the provision of customized service." Customized services "involve the sale of or long-term lease of transponders or space segment capacity for communications that are not interconnected with public-switched message networks. Customized services include intracorporate networks and television transmission. Emergency restoration services would also constitute a customized service."[30]

The process of changing U.S. international satellite communications policy reached its anticipated conclusion with the FCC's lengthy "Separate Systems" Report and Order of July 25, 1985. The FCC agreed with the White House and the Departments of Commerce and State that "separate satellite systems will provide substantial benefits to the users of international communications services."[31] The FCC also found that a rigorous enforcement of the restrictions set forth by Commerce and State on types of services that the new systems would provide, along with the legal, financial, and technical qualifications to be imposed by the FCC, would result in a system that would not cause significant harm to INTELSAT. INTELSAT, according to the FCC, "will remain the exclusive provider of satellite facilities for public-switched message services and already has a start in providing the services with which the separate satellite systems would compete. INTELSAT may lose a small part of its business in this market to new entrants, but we conclude that growing demand for satellite services will more than compensate for a smaller market share."[32]

The coup de grace to the older U.S. policy was delivered on April 3, 1986, with the FCC's Memorandum Opinion and Order in which it reaffirmed its "separate satellites decision" with certain additions. "We have determined that separate system operators should be allowed to provide occasional use television. We have also agreed to broaden our definition as to the extent to which a separate system may be used for ancillary domestic services. And we have imposed a due diligence requirement on conditionally authorized applicants to avoid warehousing of orbital slots."[33]

Seven additional U.S. companies have followed Orion's example and filed applications for separate satellite systems with the FCC. All are subject to the same American and INTELSAT coordination requirements and procedures. One system, PanAmSat, has received FCC and INTELSAT approval and has been launched.[34] Another system, Orion, has received FCC and INTELSAT approval and is still to be launched. Others are still waiting for FCC and INTELSAT approval. Many of the latter still fit the label of a "paper sat."[35]

All of these companies hope to capture a part of the international telephone services market, the revenues of which increased fourfold between 1975 and 1987, from $590,000,000 to $2.4 billion.[36]

RECENT DEVELOPMENTS

The FCC has made decisions affecting the distribution of circuits among available international facilities almost since the advent of communications satellites in 1965. The Commission's intervention in circuit distribution was a response by the Commission to its obligations under the Communications Satellite Act of 1962, which provided the framework for the creation of INTELSAT. The circuit distribution policy was instituted to ensure that AT&T and the international record carriers in the light of their vested capital investment in submarine cables "made adequate use of the global satellite system".[37] This policy enforced "loading" requirements upon AT&T, most recently involving a near 60/40 split of traffic between the use of submarine cable and the INTELSAT satellite system.

In April, 1987, the FCC issued a Notice of Proposed Rulemaking, suggesting that perhaps the time had come to eliminate loading requirements, which at that time applied only to AT&T's International Message Telephone Service (IMTS) and 800 service. Quiet government diplomacy and the fear of the effects that a sudden decision to eliminate all loading requirements would have on the two companies helped them to reach an agreement. In October, 1987, COMSAT and AT&T announced that AT&T agreed to route one-third of all of its new international traffic across COMSAT circuits (and thus INTELSAT) through the year 1994. In March, 1988, the FCC announced its decision to eliminate loading requirements. As a result of the agreement between COMSAT and AT&T the effects of the FCC decisions were cushioned for COMSAT, while AT&T gained guaranteed back-up capacity for TAT-8 and TAT-9 and the new trans-Pacific cables.[38]

In addition, the decision allowed AT&T to enjoy the less-regulated status of the other American international carriers such as US Sprint Communications Co. and MCI International Inc.[39] AT&T was for some time the only international carrier subject to loading requirements, all other carriers being able to choose whatever transmission medium they preferred. This is not to infer that there is yet adequate competition in this market, however, since AT&T is still by far the dominant carrier, holding an international market share of 92.9%, followed by MCI's 4.8% and U.S. Sprint's 1.9%.[40]

Further, the decision enabled AT&T to take full advantage of its position as part owner of several fiber optic cables and as one of the world's leading manufacturers of fiber optic cable. AT&T does not have any direct business interest in the manufacture and launch of satellite systems. Since its investment in the cable is fixed (as is that of the other owners), this means that under rate base regulation the more traffic that can be put on the cable, the lower the cost per circuit.[41] By the same token, some of the other owners of the cable are service providers based in Europe, each of which has the incentive to load up the cable so that per-circuit decreases in cost can be shared among a small pool of owners. Contrast this with the INTELSAT system, where the benefit of higher usage is shared among all participants, whether or not they have contributed to the increased traffic.

The FCC described this decision as the capstone of its policy of allowing all U.S. carriers complete freedom in choosing their modes and vendors of international transmission. The decision was also aimed at encouraging INTELSAT to develop practices suited to a more competitive market.[42]

REACTIONS TO THE U.S. POLICY OF COMPETITION

As might be expected, INTELSAT was quick to react to the change in U.S. policy. Other reactions have been slower in coming.

INTELSAT

INTELSAT is the organization with the most at stake in the evolution of U.S. international satellite policy. The emergence of other common-user satellite systems in the early 1980s has already given INTELSAT a taste of competition. The advent of private U.S. separate systems is a cause for even greater concern at INTELSAT. INTELSAT regards its operation as a potential source of "significant economic harm" and damage to its long-term viability as a provider of international telecommunications.[43]

This section gives an overview of the challenges that separate U.S. systems hold for INTELSAT. It also considers new INTELSAT services, operating procedures, and approaches that were developed as responses to the new competitive international communications environment.

The membership of other global and regional satellite communication systems is comprised of states, most of which are also members of INTELSAT and as a consequence have an interest in its continued success. Most of these states generally lack either economic incentives or capital resources to compete with INTELSAT. The private, separate system candidates in the U.S., on the other hand, need concern themselves only with their clients, and as a result have no vested interest in maintaining the global viability of INTELSAT.

Separate systems which meet the conditions of Article XIV of the INTELSAT Agreement, which specifies the procedures concerning economic harm, should gain in influence once they are in orbit. After these systems have shown themselves effective in providing service to a few initial countries, the precedent will be set. Additional affluent countries, previously resistant to the sales pitches of separate-system operators, will be likely to take advantage of an expanded array of services available at lower prices. The net effect of any one separate system might be negligible, but a host of separate systems could drastically alter the entire market structure and psychology of the international telecommunications market.

INTELSAT is particularly vulnerable to such a shift in market structure. Since the beginning, the INTELSAT system has employed price-averaging so that developing countries could be served economically. This results in a cross-subsidy situation, akin to that experienced in the U.S. prior to divestiture, where business long-distance users subsidized expenses for local service provided to residential users. INTELSAT, in its effort to provide "universal service" to all member countries, fixed the price for any given service independently of the volume of traffic over the route provided by the service. The fact that the separate system applicants all come from the U.S. is especially disquieting to INTELSAT. Despite its global coverage, INTELSAT is vulnerable to a fundamental reliance on a particular region, the North Atlantic, which accounts for 40% of its traffic.[44] This revenue provides a substantial subsidy for many of the other regions served by INTELSAT, as the economies of scale along the Atlantic route make the unit cost for its services lower than for other regions with lighter traffic loads.[45]

The exceptions to INTELSAT's price averaging involve even clearer subsidies. An example of this is Caribnet, which offers 50% discounts on INTELSAT Business Service and Intelnet Service to users in the Caribbean. Faced with competition in a portion of its service area where price-averaging makes prices higher than they would otherwise be, INTELSAT now finds itself at a pricing disadvantage relative to alternative systems and technologies.

However, if INTELSAT moves away from price-averaging, it is

questionable whether those users on its lighter traffic routes, often poor developing countries, will be able to pay the resulting higher rates. Compromising the universal service concept strikes at the very heart of INTELSAT's overall purpose of providing a global network available to all countries on an equitable basis.

In 1986 the INTELSAT Board of Governors affirmed that future planning decisions would have to assume resource-based tariffs, and in 1987 it agreed to a process for implementing such tariffs. The move to resource-based tariffs followed recommendations of INTELSAT's Charging Policy Working Group and a request of the signatories to find ways to motivate long-term commitments to the INTELSAT system as well as stimulate the generation of new traffic. INTELSAT regards resource-based pricing as a means to optimize system use and promote "economic efficiency to sustain INTELSAT in the current and foreseeable industry environment."[46]

Additional factors, such as the U.S. removal of loading requirements, the quickly-changing pace of technology, and the lucrative incentives which attract new carriers into the long-distance telecommunications market for both satellite and submarine fiber-optic cable transmissions, all put the original objectives of INTELSAT and its ability to meet them into question. INTELSAT policy is as much affected by the *threat* of competition as by actual competition. Trends in international traffic patterns are also of concern to INTELSAT. While voice service accounts for only 69% of its transmissions, compared to 90% for telephone traffic in general, non-voice services may dominate the offerings of the separate systems. They hope to tap into a data services market which some analysts believe will ultimately account for half of all telephone traffic. Clearly if INTELSAT wants to maintain a competitive revenue base, it will have to offer more data and video services.[47]

INTELSAT is not lacking in interested support, though. Many PTTs cross-subsidize other government programs with the profits derived from the services they provide in conjunction with INTELSAT. The price that a PTT pays for the total cost to access INTELSAT capacity must be placed in perspective; it represents less than 10% of the total long-distance charge billed to users.[48] The first reaction of a foreign PTT to a separate system would probably be to consider the savings that it might offer to users as insignificant and not worth the consequent difficulties of coordination with INTELSAT and the additional domestic regulation and loss of revenue. This, along with INTELSAT's reputation for reliable service and its impressive track record in the industry, assures continued PTT support for INTELSAT. Furthermore, the PTT will likely have made a capital investment in INTELSAT and related earth structure facilities. Even if a PTT gives landing rights to a separate system, access to its communications facilities may come with such an array of restrictions and conditions that the separate system will offer no real threat to INTELSAT. So if the first orbiting separate systems represent a potential domino effect, whereby one PTT after another approves their operation, that effect could prove to be minimal. Combating the vested interests of PTTs is the chief difficulty facing private systems, but clearly one where INTELSAT holds the main advantage.

INTELSAT also has some luxury in the protection promised by the U.S. government. The Foreign Relations Authorization Act for the fiscal years of 1986 and 1987 states that if the federal government decides to support the implementation of a separate system, the Secretary of State must submit a report to Congress outlining the foreign policy reasons for the President's determination, as well as a plan for minimizing any negative effects it may have on INTELSAT and on U.S. foreign policy interests.[49]

In attempts to become more competitive and assuage doubts about its ability to become a more efficient and market-driven entity, INTELSAT has offered new services, lowered costs, and claimed to have increased its efficiency. INTELSAT Business Service and the many of the other services mentioned in this chapter and Chapter Two are examples of INTELSAT's efforts to become competitive in the area of business communications. In introducing these new services, INTELSAT was attempting to show that the separate system satellite applications promised no services which INTELSAT did not already provide or plan to provide in the near future. INTELSAT has indicated that its attempt to develop the IBS service was the stimulus for the first separate satellite system proposals by Orion and ISI. This perception was outlined in part in the "Petition To Deny Application of Communications Satellite Corporation," filed on April 5, 1983, with the FCC in response to the Orion application. The separate system applicants countered that INTELSAT offered IBS only because of the competitive threat which separate systems represented. However, INTELSAT Business Service and other new INTELSAT offerings were in the planning stages before the separate systems made their applications with the FCC. Nevertheless, INTELSAT has acknowledged that its approval of the services may have been accelerated by the anxiety that its member-nations felt about the prospect of separate, competing systems.[50]

INTELSAT's boasts of increased efficiency and lower costs to lease half-circuits should be seen in the light of the fact that the price for technological goods and services usually declines once the initial cost of research and development has been recovered. The initial drop of $12,000 to lease a half-circuit (from $32,000 to $20,000) occurred in the first six months of 1965, suggesting a previous overestimate of cost and expenses.[51]

Leland B. Johnson, a senior economist at the RAND Corporation, believes that "INTELSAT's global cost-sharing arrangements encourage countries to over-estimate their individual traffic needs, for they are billed for costs only in proportion to their actual use of INTELSAT facilities," therefore contributing to the excess capacity problem facing INTELSAT and the industry as a whole.[52] Johnson also argues that INTELSAT's investment policies are inappropriate given the organization's forecasting trends, as well as the technological advancement of fiber optics and the advent of the separate satellite systems. INTELSAT overestimated its future traffic by 10% in 1985, while concurrently ordering five additional VI satellites priced at $232 million each.[53]

With a growing interest by individual countries in establishing their own systems, genuine competition from alternative transmission technologies, and an uncertain regulatory environment within the United States, INTELSAT's future promises to be full of technical, economic, and political challenges.

Foreign Reactions

The hope and expectation that the U.S. policy of encouraging competition in international satellite communications would stimulate other nations to adopt policies that would make it possible for American satellite systems to enter foreign markets is clearly stated in the following excerpt from *NTIA Telecom 2000*:[54]

> Changes in U.S. communications and information services markets are reflected in some of the changes which have occurred and are occurring in other countries. These changes are due, in part, to the fact that many of the forces for change manifest here--new technology, demographic changes, and greater emphasis on efficiencies believed to result from more reliance on these "electronic tools," for example--are increasingly universal phenomena. They also reflect marketplace pressures to conform with changes occurring in the world's single largest communications market, the United States. The dynamic effects of U.S. domestic deregulation and competition, in short, are themselves having an international impact and driving many other countries in the world toward liberalization, *i.e.*, more procompetitive policies.

The United States maintains that it is not attempting to impose its policies on other countries. However, American policy-makers hope that other countries will realize that competition is in their best interests. Such is the theme of the following excerpt from *NTIA Telecom 2000*:[55]

> We cannot and should not unilaterally impose our vision upon the world, even if it derives from basic values or successful domestic results. Nonetheless, with vigorous advocacy, we have seen nations accept U.S. policy initiatives, for example regarding the benefits of competition from international satellite systems separate from INTELSAT, and other similar approaches which already have enhanced consumer welfare in the United States. We expect more countries to rely on competition and open markets, ultimately because such procompetitive policies are in their interests as well as our own, and because failure to [adopt] such approaches, with respect to communications particularly, will be costly.

Unfortunately, international cooperation has been minimal because of several factors. The biggest hurdle facing separate system applicants has been in obtaining approval from foreign PTTs. The foreign PTTs are primarily concerned that separate systems will make it more difficult for domestic programs to remain profitable. As a result, there is no perceived benefit from using a U.S. company to provide communications services.

Obviously, any international call involves two different telecommunications providers: one in the country where the call originates, and one in the country where the call terminates. In the North Atlantic route, non-U.S. providers in Europe are usually PTTs. Specific interests and biases toward either satellites or fiber optic cable vary with PTTs, but they cannot be ignored. In Europe, the routing of traffic via cable is important to countries that collect transit charges from call-termination countries, countries lacking their own cable landing points. European PTTs generally control both satellite and cable facilities, whereas in the United States the two are in different hands.[56] No matter what the particular circumstances or point of view, the important point to remember is that the perspective and motivations of the PTTs must be considered. The success of an FCC regulatory decision may be heavily influenced by interests outside of the United States.

According to Henry Rivera:[57]

One of the biggest challenges for separate systems has been obtaining approval from foreign PTTs. Generally, the PTTs fear that additional competition will reduce their revenue and thereby make it more difficult for them to subsidize their domestic postal and telephone service. They think that dealing with separate systems will increase their transaction costs without providing any significant benefits. Many PTTs believe that the volume of international traffic will remain constant regardless of the number of U.S. carriers providing service. In some cases nationalism is also a factor--the foreign governments are reluctant to give U.S. companies a stronger foothold in providing communication service.

There has been activity in certain quarters, however, that may give rise to a guarded optimism. The European Green Paper, for instance, has been hailed as an example of future liberalization. Although the Green Paper is concerned primarily with the creation of a single internal common market for telecommunications by 1992 and emphasizes the opening up of the European terminal market quickly and the liberalization of the market for telecommunications services no later than 1989, it is believed that it will at least set the stage for competition in satellite services.[58] As the European suppliers of satellite services both government and private grow strong and confident, it is possible that they will become willing to meet the U.S. private satellite providers head-on with the hope of giving them real competition.

A more important indication of a possible change in policy on the part of European countries was the decision of the United Kingdom and West Germany to vote favorably in INTELSAT's Board of Governors on the convening of an Extraordinary Meeting of Signatories to consider PanAmSat's application to build a separate satellite system.[59] An even clearer indication would be a decision on the part of the United Kingdom and West Germany, or other important European powers, to join with a private satellite system such as PanAmSat in providing services on the potentially lucrative North America-Europe route. The decision of British Aerospace to become an active partner in the Orion enterprise makes this possibility seem even more likely.

CONCERN OVER FIBER OPTICS

As discussed in Chapter One, fiber optic cable is probably causing the most concern to the INTELSAT and international satellite communications industry. Despite the many private satellite networks across the country, satellites carry a small portion of overall domestic long-distance telecommunications traffic in the United States. Recent trends indicate a shifting of much of that traffic away from satellites and older systems towards the increasing use of fiber optics. Causing much greater concern among potential international satellite communication operators, though, is the growing competitive threat presented by fiber optics in the provision of point-to-point service along transoceanic routes. These routes include the heavily-used North Atlantic route between Europe and the United States, from which INTELSAT derives about 40% of its revenues. Many industry observers predict that fiber optic cables will become the major service providers along high-volume oceanic routes. They point to fiber's technical superiority, citing its imperceptible transmission delay, and fiber's economic advantages in serving high-volume routes, as reasons for this trend.

Whereas the replacement of coaxial cable or terrestrial microwave with optic fibers is perceived as a natural system upgrade implementing the latest and most effective technology, the prospect of replacing satellite systems with optic fibers elicits more attention and emotion. Rarely does one find a magazine article or conference paper which discusses the impact optical fibers are having on the copper industry, but there are countless articles addressed to the issue of "fiber optics vs. satellites." As an example, the *Proceedings of Fibersat 86*, the 1986 International Conference on Satellite and Fiber Optic Communications, contains more than 90 papers concerning the merits of fiber or satellite transmission.[60]

Two major reasons explain why the satellite vs. fiber debate has commanded so much attention. First of all, the issue is more than just a technical debate; it has political ramifications. As mentioned above, the TAT-8 Atlantic Ocean cable system is positioned to compete aggressively along INTELSAT's most lucrative route. Because INTELSAT is a political creature, a government-run global organization whose members assign varying and qualified priorities to its success, international politics and perceived national self-interests will play a decisive role in the outcome of the fiber-satellite debate.

Secondly, both fiber optic cables and the newer satellites have the capacity to handle very high data rates and large volumes of traffic. These two transmission methods will be the main "communication highways" of the future. If current projections of large international traffic increases are accurate and a distinct victor in the fiber-satellite competition emerges, the victor should enjoy superior economies of scale sufficient to drive the other to marginality if not out of business.

SUMMARY CONCLUSIONS

The U.S. policy favoring the concept of a universal monopoly of international satellite communications through the auspices of INTELSAT was bound to fail. As was indicated in the preceding chapter, as more countries acquired the necessary technology there arose the possibility of a series of international and regional competitors. In addition, domestic developments were working to defeat the concept. American entrepreneurs saw profits to be made in specialized systems, both domestic and international, and their prospective clients began to demand the services that such companies were hoping to offer. There was also a desire to avoid the rather high tariff rates of the PTTs.

The key development in the move away from the global monopoly concept was the increased emphasis on free enterprise that marked the Nixon and Reagan administrations. Individuals and groups desiring to enter the field of satellite communication found particularly receptive ears in Nixon and Reagan and some of the central figures in their respective bureaucracies. Competition in domestic and international satellite communication became feasible shortly after the first filing of applications with the Federal Communication Commission. This was soon followed by additional procompetitive policy moves such as the threat to eliminate the cable-satellite loading requirements.

The reaction to the new U.S. policy of competition has been one of increased activity on the part of INTELSAT, being the most immediately affected, and of caution on the part of those countries which the Americans hoped would also adopt market-oriented policies. INTELSAT, for example, quickly made a number of changes in its policies that it believed would place INTELSAT in a more competitive position with prospective U.S. international satellite communication service providers. Few of the European countries on the other end of the heavy traffic America-Europe corridor have emulated the U.S. either by encouraging their domestic communication enterprises to enter the race for international markets or by opening their domestic markets completely to existing or potential U.S. satellite communication service companies.

All of this could change, however. As Europe proceeds further with its efforts at economic integration it might gain the confidence necessary to challenge U.S. satellite communication companies at their own game. Individual countries such as the United Kingdom might adopt a pro-competitive stance similar to that of the United States. The PanAmSat and Orion stories may indicate that some of the first moves in that direction have already been made.

89

NOTES

1. See the account by Marcellus S. Snow, "Natural Monopoly in INTELSAT: Cost Estimation and Policy Implications for the Separate Systems Issue," *Telematics and Informatics*, Vol. 4, No. 2 (1987), p. 136.

2. Robert S. Magnant, *Domestic Satellite: An FCC Giant Step* (Boulder, CO: Westview Press, 1977), pp. 64-65.

3. U.S., Public Law 87-624, *Communications Satellite Act of 1962*, Sec. 102. (a).

4. *Ibid.*, Sec. 102. (d).

5. As quoted in Magnant, pp. 92-93.

6. For an interesting discussion of the Bell system at this time and how these decisions affected it, see Jeremy Tunstall, *Communications Deregulation: Unleashing America's Communications Industry* (New York, N.Y: Basil Blackwell, 1986), Chapter 9.

7. See Magnant, pp. 146-149 and *NTIA Telecom 2000*, Chapter I, "Twenty Years After The Rostow Report."

8. As quoted in Magnant, pp. 156-157.

9. *Ibid.*, p. 171.

10. FCC, Second Report and Order, June 16, 1972, In the Matter of Establishment of Domestic Communications-Satellite Facilities by Non-governmental Entities, Docket No. 16495 (35 FCC 2d 844) p. 845.

11. *Ibid.*, p. 847.

12. *Ibid.*, p. 846.

13. In the Matter of Domestic Transponder Sales, *90 FCC 2d 1238*, (1982).

14. *Ibid.*, pp. 1246-1248.

15. As quoted in Rob Stoddard, "Decade of Deregulation," *Satellite Communications*, July 1987, pp. 21-22.

16. *NTIA Telecom 2000*, p. 274.

17. John E. Koehler, "Satellite Communications, International Considerations," *IEEE Communications Magazine*, Jan. 1987, pp. 33-35.

18. "U.S. Domsat Fleet Strengthened as FCC Recognizes Operator Replacement Rights," *Satellite Communication*, Jan. 1989, p. 8.

19. As quoted in *Ibid.*

20. "Consortium Mandated for Mobilesats," *Satellite Communications*, Feb. 1987, p. 10.

21. See Guy M. Stephens, "MSS Market Heats Up," *Satellite Communications*, June 1988, p. 12 and Michael A. Dornheim, "Potential Satellite Services Laboring Under Conflicting Frequency Schemes," *Aviation Week and Space Technology*, Jan. 4, 1988, pp. 47-48. North American Mobile Satellite Corp. is a Class B member of the consortium and has no voting rights. The other seven companies are Class A members, enjoying voting rights.

22. See Peter H. Cowhey and Jonathon D. Aronson, "The Great Satellite Shootout," *Regulation*, May/June 1985, pp. 29-31.

23. Cowhey and Aronson, p. 33.

24. U.S., Department of State and Department of Commerce, Senior Interagency Group on International Communication and Information Policy, *A White Paper on New International Satellite Systems*, Washington, D.C., February 8, 1985. The President's announcement can be found in Appendix A.

25. *Ibid.*, p. 43.

26. *Ibid.*

27. *Ibid.*, p. 44.

28. *Ibid.*, p. 26.

90

29. *Ibid.*, p. 30.
30. *Ibid.*
31. *FCC 85-399, 36029*, p. 5.
32. *Ibid.*
33. FCC, *Memorandum Opinion and Order, April 3, 1986*, p. 33.
34. "Summer's Events Enable PanAmSat to Shed Image of Being 'Paper Sat'," *Satellite Communications*, Aug. 1988, p. 8. See also below, Chapter Five.
35. "Orion Finally Wins INTELSAT Clearance, *Satellite Communications*, Sept. 1989, pp. 13-14.
36. "FCC's 1975-87 Trends Report Shows U.S. Overseas Telephone Carrier Revenues Grew 300%," *Telecommunications Reports*, Oct. 24, 1988, pp. 23-24.
37. *Ibid.*
38. See Charles Mason, "Circuit Loading Guidelines Out," *Telephony*, Mar. 28, 1988, p. 5 and Elizabeth Tucker, "COMSAT Signs Contract to Ensure AT&T Traffic," *Washington Post*, Oct. 10, 1987.
39. Both MCI and Sprint, it should be noted, lease transportation time on INTELSAT satellites.
40. "FCC's 1975-1987 Trends Report Shows U.S. Overseas Telephone Carrier Revenues Grew 300%," *Telecommunications Reports*, Oct. 4, 1988, pp. 23-24.
41. As a result of the new "price cap" regulation this has become less important.
42. Mason, p. 5.
43. See "Authorization of Private International Satellite Systems . . . ," p. 28.
44. *INTELSAT Report, 1987-88*, p. 39.
45. Sigrid Arlene Mendel, "Authorization of Private International Satellite Systems in Competition with COMSAT: An Analysis of the Underlying Legal Justification and Policy Factors," *Law and Policy in International Business*, Vol. 18, 1986, p. 285.
46. INTELSAT, *Annual Report by the Board of Directors, 1988*, p. 18-19.
47. *INTELSAT Report, 1987-88*, p. 13-14.
48. Leland L. Johnson, "Excess capacity in international communications," *Telecommunications Policy*, Sept. 1987, p. 283.
49. *Foreign Relations Authorization Act, Fiscal Years 1986 and 1987*, Pub. L. No. 99-93, 146 (d), 99 Stat. 405, 425-26 (1985).
50. Tom Kerver, "North Atlantic Fury," *Satellite Communications*, Aug. 1985, p. 24.
51. *INTELSAT Annual Report, 1983*, p. 31.
52. Leland L. Johnson, "Excess Capacity . . ., pp. 283-286.
53. *Ibid.*
54. *NTIA Telecom 2000*, p. 41.
55. *Ibid.*, p. 119.
56. *Ibid.*
57. See Henry M. Rivera, "Separate Systems and International Communications," *IEEE Communications Magazine*, Vol. 25, No. 1 (Jan. 1987), p. 40.
58. See OECD, *Trends of Change in Telecommunications Policy*, Paris, 1987. For a commentary see Karl-Heinz Narjes, "Towards a European Telecommunications Community: Implementing the Green Paper," *Telecommunications Policy*, June, 1988, pp. 106-108.
59. See below, Chapter Five, The Case of PanAmSat.
60. See *Conference Proceedings, Fibersat 86*.

CHAPTER FOUR

INTERNATIONAL ALLOCATION OF

SPACE RESOURCES

INTRODUCTION

In order for satellite communication to be an integral part of the
international telecommunication network, those who intend to operate
satellite communication systems must have access to the necessary
space resources. Space resources in this context include the
appropriate bands of radio frequencies for uplinks and downlinks and
the appropriate geostationary orbital positions on which to situate a
communication satellite. In addition, once the satellite system is in
operation, or well into the planning stage, the operator must be able
to proceed with the knowledge that other systems will not create
harmful interference to his operation.

In the early days of satellite communication it appeared as though
the available supply of space resources would be more than adequate
to meet the needs of all those who wished to participate. The fervor
with which some of the more powerful developed countries embraced
this new technology for communication and other national purposes,
however, led some nations to fear that when the time came for them
to take advantage of this new technology all of the more desirable
space resources would already have been appropriated, thus forcing
them to accept lesser and more expensive alternatives. These
countries fought for and obtained, in certain parts of the existing
allocation system, changes that they felt would guarantee them equal
access to the necessary resources when they needed them.

The purpose of this chapter is to present the procedures and the
attendant problems that are involved in obtaining these resources and

the potential impact that changes therein can have on the future of satellite communications. There are two components to this process: the domestic and the international. The domestic component, described in the next chapter, is governed to a great extent by what is decided at the international level, the subject of this chapter.

THE ROLE OF THE INTERNATIONAL TELECOMMUNICATION UNION

Very early in the history of radio it became apparent that international cooperation was necessary in order for radio communication to fulfill its true potential. The countries using radio at that time created an unofficial international organization, known as the International Radiotelegraph Union, to draw up the necessary international standards and regulations.[1] In 1936 the International Radiotelegraph Union merged with the older International Telegraph Union to create the International Telecommunication Union (ITU). When communication by satellite became feasible, the ITU took over the role of providing a forum for countries to draw up the standards and regulations necessary for this new technology to be incorporated into the world's telecommunication system.

The ITU is made up of eight major components, plenipotentiary conferences, administrative conferences, an Administrative Council, and five permanent organs: the International Frequency Registration Board (IFRB); the International Radio Consultative Committee (CCIR); the International Telegraph and Telephone Consultative Committee (CCITT); the Bureau for Telecommunications Development; and a General Secretariat.[2]

For our purposes the most important ITU organs are the administrative conferences and the five-member elected International Frequency Registration Board. It is the ITU international administrative radio conference, in which each member of the 167 members of the ITU has an equal vote, that establishes the Radio Regulations containing the rules for gaining access to space resources and the process by which a country gains the right to use those resources without the threat of harmful interference from other systems.

Probably the most important change in the ITU since it was created in 1934, and one which has a strong bearing on our subject, has been its evolution from an organization dominated by a few developed countries to one in which the vast majority of its members are developing countries with different interests and needs in the field of telecommunications. The effects of this change will become apparent as we proceed.

ALLOCATION OF RADIO FREQUENCIES
TO SATELLITE SYSTEMS

One of the more important functions of the ITU is the maintenance
of the Table of Frequency Allocations. The Table, which takes up a
major portion of the Radio Regulations, is a record of the allocation
of discrete portions of the usable radio frequency spectrum to specific
types of radio services made by administrative radio conferences over
the years.[3] The 1982 edition covers frequencies from 9 kHz to 400
GHz. Several dozen different uses of radio, including satellite
communication, have been given allocations of frequencies in that
Table.[4]

The use to which the Frequency Allocation Table is put depends
upon a number of factors, including historical usage and politics. For
example, the Berlin Radiotelegraph Conference of 1906 allocated the
frequencies 500 and 1000 kHz to public correspondence in the
maritime service and frequencies below 188 kHz for long-distance
communication by coast stations. Frequencies between 188 and 500
kHz were reserved for mobile "services not open to public
correspondence," in other words military services.[5] The decision by
that conference to allocate these frequencies to naval communication
was dictated primarily by the fact that early radio used frequencies in
this range and they were found to be especially useful by the navies of
that day. The maritime service still lays claim to many of the
frequencies in these areas.

As new radio services came into being, they requested and were
given discrete portions of the usable radio frequency spectrum for
their own use. As the radio frequency spectrum became more and
more crowded with services despite the expansion of the usable
frequency range, the process of adjusting the Table of Frequency
Allocations became increasingly complicated. As a result,
administrative radio conferences with the authority to revise the Table
became the scene of serious confrontations. Services began to
compete actively with each other for frequency bands or portions of
frequency bands, and countries that tended to use one service in
preference to another entered the fray. Further, government needs,
including military considerations, have continued to play an important
role in the allocation procedure. As the process is described by
Anthony M. Rutkowski:[6]

There are often important technical reasons for making these
allocations. There are also safety, marketing, and scientific
considerations which dictate certain results. However, the bases
for many, if not most, of the allocations lie in domestic and

international political-economics. Various user groups find the allocation process a convenient means of securing a guaranteed monopoly of use or an advantage for themselves. Intense lobbying usually accompanies domestic and international allocation proceedings.

The Allocation Table is complicated by a number of practices that have evolved over the years. One such practice is to permit more than one radio service to operate in the same frequency band. For instance, the 14.47-14.5 GHz frequency band is shared by the fixed radio, fixed satellite (Earth-to-space), mobile (except aeronautical mobile), and radio astronomy services. Another practice is to give different levels of priorities to services. In the example given above, for instance, the fixed, fixed satellite, and mobile services are all classified as "primary" and thus have equal rights to use the frequency band in question. The radio astronomy service, however, is classed as "secondary" and thus cannot cause harmful interference to stations of the other services; nor can it claim protection from harmful interference from those stations. In addition, in those frequency bands where it is feasible, different allocations in the same frequency band are made for different regions of the earth. For instance, in Regions 1 and 3 the 14.3-14.4 GHz frequency band can be used for fixed, fixed satellite (Earth-to space), mobile (except aeronautical mobile) services on a primary basis, and the radionavigation satellite services on a secondary basis, while in Region 2, the Americas, it may be used only for the fixed satellite service (Earth-to-space) on a primary basis, and the radionavigation satellite services on a secondary basis.[7]

One final practice that makes it difficult to know exactly how any specific frequency band is being used is that which allows individual countries or groups of countries to deviate from the overall allocation scheme. These deviations, which are extensive, are indicated by a numerical reference contained in the band in question and are known as "footnotes." For example, the 13.4-14 GHz band is allocated throughout the world to the radiolocation service on a primary basis and to the standard frequency and time signal satellite (Earth-to-space) and space research services on a secondary basis. However, according to footnote 853, this band and the band 13.25-13.4 GHz are also allocated to the fixed service on a primary basis in Bangladesh, India, and Pakistan. According to footnote 854, thirty-nine other countries also allocate the band 13.4-14 GHz to fixed and mobile services on a primary basis, and according to footnote 855 eleven other countries also allocate it to the radio navigation service on a primary basis.[8]

The process of obtaining adequate frequency allocations for satellite communications has always been a complicated one. The first ITU conference to become involved in satellite communication was the 1959 Administrative Radio Conference. Unsure of just how the new service would evolve, the conference allocated the following nine frequency bands to "space" and "Earth-to-space" for research purposes. All bands were to be shared with other radio services:[9]

1) 136-137 MHz. Space and Earth-to-space, shared with Fixed and Mobile as primary services.

2) 400-401 MHz. Space and Earth-to-space, shared with Meteorological Aids as primary services.

3) 1 427-1 429 MHz. Space and Earth-to-space, shared with Fixed and Mobile except aeronautical mobile as primary services.

4) 1 700-1 710 MHz. In Region 1 the Fixed service was the primary service, while Space, Earth-to-space, and Mobile were secondary services. In Regions 2 and 3, Fixed and Mobile were the primary services while Space and Earth-to-space were secondary services.

5) 2 290-2 300 MHz. In Region 1 the Fixed service was the primary service and Space, Earth-to-space, and Mobile were secondary services. In Regions 2 and 3, Fixed and Mobile were the primary services and Space and Earth-to-space were secondary services.

6) 5 250-5 255 MHz. Radiolocation was a primary service and Space and Earth-to-space were secondary services.

7) 8 400-8 500 MHz. Fixed and Mobile were primary services and Space and Earth-to-space were secondary.

8) 15.15-15.25 GHz. Space and Earth-to-space were primary and fixed and mobile were secondary services.

9) 31.5-31.8 GHz. Space and Earth-to-space were primary services and fixed and mobile were secondary.

By the time the 1963 ITU Extraordinary Administrative Radio Conference was convened, the world's knowledge of space communication had progressed to the point where it was possible for the majority of delegations, including both space powers, to agree to allocate two frequency bands for the exclusive use of space communications: 7 250-7 300 MHz and 7 975-8 025 MHz. All other services were asked to remove themselves from these bands by January 1, 1969. Approximately twenty delegations, however, were unable to accept the decision, and via the footnote process decided that it was necessary for them to continue to give the fixed and mobile services, as well as the space communication service, a primary status in those bands.[10]

In addition, the 1963 conference allocated the following frequencies to space communication on a shared basis with other services:[11]

1) 3 400-3 700 MHz.
2) 3 700-4 200 Mhz.
3) 4 440-4 700 MHz.
4) 5 725-6 425 MHz. in Region 1, 5 925-6 425 MHz. in Region 2, and 5 850-6 425 MHz. in Region 3.
5) 7 300-7 750 MHz.
6) 7 900-7 975 MHz.
7) 8 025-8 400 MHz.

The 1971 ITU World Administrative Radio Conference for Space Telecommunication added approximately 177 GHz of radio frequency spectrum for space communications services, all of which were to be shared with other terrestrial radio services.

By the time of the 1979 ITU World Administrative Radio Conference, the various uses for satellites were fairly well sorted out. After recognizing mobile satellite communications as a new service, the conference set aside additional frequencies for satellite communications, especially in bands above 40 GHz. The evolution in satellite communications making the addition of more frequencies necessary is explained in the following manner in the report by Chairman of the U.S. delegation to that conference:[12]

The U.S. proposed that specific allocations be made for both space and terrestrial services in the bands from 40 GHz to 300 GHz. Specific allocations were necessary for several reasons. First, many spectral lines had been identified which are important to the passive radio services--radioastronomy, space research and earth exploration satellite. The unique physical properties of the lines dictate the allocations for these services. Second, development of equipment for frequencies above 40 GHz was currently in progress. Specific allocations would provide guidance for the orderly development of new technology and the next generation of telecommunications systems. Finally, it was not certain when the next radio conference competent to deal with the spectrum above 40 GHz would be convened. In view of the rapid development of the recent past, it was felt necessary to prepare for the future by providing sufficient allocated bands in a flexible arrangement to accommodate the foreseen requirements.

The 1979 allocations above 40 GHz for satellite communications (broadcasting, fixed, mobile, and amateur) were as follows:[13]

1) 40.5-42.5 GHz., Broadcasting Satellite, primary, shared.
2) 42.5-43.5 GHz., Fixed Satellite (Earth-to-space), primary, shared.
3) 43.5-47 GHz., Mobile Satellite, primary, shared.
4) 47-47.2 GHz., Amateur Satellite, primary, shared.
5) 47.2-50.2 GHz., Fixed Satellite (Earth-to-space), primary, shared.
6) 50.4-51.4 GHz., Fixed Satellite (Earth-to-space) on a primary basis and Mobile Satellite (Earth-to-space) on a secondary basis, shared.
7) 66-71 GHz., Mobile Satellite, primary, shared.
8) 71-74 GHz., Fixed Satellite (Earth-to-space) primary and Mobile Satellite (Earth-to-space), primary, shared.
9) 74-75.5 GHz., Fixed Satellite (Earth-to-space), primary, shared.
10) 75.5-76 GHz., Amateur Satellite, primary, shared.
11) 76-81 GHz., Amateur Satellite on a secondary basis, shared.
12) 81-84 GHz., Fixed Satellite (space-to-Earth) primary, and Mobile Satellite (space-to-Earth), shared.
13) 84-86 GHz., Broadcasting Satellite, primary, shared.
14) 92-95 GHz., Fixed Satellite (Earth-to-space), primary, shared.
15) 95-100 GHz., Mobile Satellite, primary, shared.
16) 102-105 GHz., Fixed Satellite (space-to-Earth), primary, shared.
17) 134-142 GHz., Mobile Satellite, primary, shared.
18) 142-144 GHz., Amateur Satellite, primary, shared.
19) 144-149 GHz., Amateur Satellite on a secondary basis, shared.
20) 149-150 GHz., Fixed Satellite (space-to-Earth), primary, shared.
21) 150-151 GHz., Fixed Satellite (space-to-Earth), primary, shared.
22) 151-164 GHz., Fixed Satellite (space-to-Earth), primary, shared.
23) 190-200 GHz., Mobile Satellite, primary, shared.
24) 202-217 GHz., Fixed Satellite (Earth-to-space), primary, shared.
25) 231-235 GHz., Fixed Satellite (space-to-Earth), primary, shared.
26) 235-238 GHz., Fixed Satellite (space-to-Earth), primary, shared.
27) 238-241 GHz., Fixed Satellite (space-to-Earth), primary, shared.

28) 241-248 GHz., Amateur Satellite on a secondary basis.
29) 248-250 GHz., Amateur Satellite, primary, shared.
30) 252-265 GHz., Mobile Satellite, primary, shared.
31) 265-275 GHz., Fixed Satellite (Earth-to-space), primary, shared.

If satellite communication continues to grow along with other radio uses, those interests involved in satellites, whether government or private, will find it necessary to assert all of the influence they can muster in future ITU conferences in order to continue to have the allocations they need in the ITU's Frequency Allocation Table.

OBTAINING THE RIGHT TO USE SPACE RESOURCES

Making certain that the necessary bands of radio frequencies are allocated to the various satellite communication services is only one of the problems facing those who would provide a satellite communication system. Another problem, just as important, is that of obtaining the right to use frequencies in the allocated bands and the accompanying geostationary orbital position without suffering harmful interference from other similar systems.

For the entity, governmental or private, that intends to provide a satellite communication system the process has two parts. First, the right to use the space resources in question must be recognized and approved by an agency of the user's own government. Second, the use of the space resource must be recognized by an agency of the International Telecommunication Union, the International Frequency Registration Board (IFRB). The international component of that process will be discussed in this chapter.

Two different international procedures for obtaining this right to use certain frequencies and the associated geostationary satellite orbital position have evolved, dubbed the first-come, first-served procedure and the *a priori* procedure.

The First-Come, First-Served Approach

Prior to the advent of satellite communications the International Telecommunication Union relied primarily on the first-come, first-served procedure. According to this process, national telecommunication administrations notified the ITU's International Frequency Board of frequencies to be used by domestic radio stations if: 1) the frequencies in question were capable of causing harmful interference to any other radio service; 2) the frequencies were to be used for international communication; or 3) the administration desired "to obtain international recognition of the use of the frequency." The IFRB examined each notification, which included technical specifications, for its conformity with the ITU Convention and Radio Regulations, and for the possibility that it would cause harmful interference to a station whose frequency had already been registered by the IFRB in the Master International Frequency Register. If the findings of the Board were satisfactory, the Board proceeded to enter the frequency in question in the Master Register, and the date of that registration determined the rights of that radio station against those of other radio stations. In the event of an unfavorable finding, there were a number of provisional and appellate procedures that could result in an eventual entry of that notification in the Master Register with full or secondary rights.[14]

The 1963 Extraordinary Administrative Radio Conference, the first ITU conference to be dedicated solely to satellite communication, decided that a similar notification and recording procedure should be used. The characteristics of a proposed satellite system, including the frequencies to be used, would be notified to the IFRB. The IFRB, in its turn, would examine the notification for conformity with the provisions of the ITU Convention and Radio Regulations, and for the possibility that it could cause harmful interference with other satellite systems. If the results were favorable, the system would be registered in the Master International Frequency Register and protected from harmful interference from new satellite systems.

The delegates to the 1971 World Administrative Radio Conference for Space Communications discussed the notification and registration procedures at great length and came to the conclusion that they were not adequate for the new geostationary satellite communication systems. At the conclusion of the debate it was decided to introduce a modified procedure containing three major components, advance publication, final coordination, and recording in the Master Register.

The advance publication process mandates that administrations planning a satellite system notify the IFRB of that fact, along with providing information concerning the system's main characteristics--including orbital positions in the case of geostationary satellite communication systems--between five and two years in advance of bringing that system into service.[15] This information is made available to other administrations by the IFRB. If an administration feels that the new system might cause harmful interference with one of its existing or planned satellite systems, it sends all pertinent information to the notifying administration, along with a copy to the IFRB, within four months of receiving the notice of the planned new system. All administrations involved are asked to work out any differences they may have including, if necessary, changing orbital positions and frequencies. The IFRB keeps the members of the ITU informed of the progress of the negotiations and provides assistance to the parties in reaching agreement if so requested.[16]

The second phase of the process begins just prior to asking the IFRB to register the satellite network in the Master Register. In this phase, the administration intending to send a final, formal notification to the IFRB of a frequency assignment for a space network is required to "effect coordination" of that assignment with any other administration that believes that the frequencies to be used will affect their own frequency assignments. If coordination is requested by another administration, all of those involved are required to exchange detailed information about their systems through the intermediary of the IFRB and to try to work out their differences. If they are not successful, the IFRB becomes automatically involved in the process. This can include a request by the Board to the parties to adopt a compromise, or an assessment by the Board of the actual interference, if any, that the new system might cause to other satellite systems. As is the case of advanced notification, the resolution of any potential conflict rests entirely on the good faith of the administrations concerned.[17]

The third and final phase involves the notification of the new satellite system to the IFRB for registration in the Master Frequency Register. As is the case with frequencies to be used in terrestrial radio services, the notification is examined by the IFRB for conformity with the ITU Convention; the Radio Regulations, including the Frequency Allocation Table conformity with the coordination procedures and the possibility that the new system might cause harmful interference to the satellite communication networks of other administrations. If a notification receives a favorable finding, it is entered in the Master Register as of the date of notification.[18] Once registered, the satellite system in question is protected from interference from any satellite systems subsequently notified to the Board.[19]

As we will see, dissatisfaction with certain aspects of the first-come, first-served procedure for both terrestrial and satellite communications resulted in the adoption of the *a priori* procedure for certain types of communication services.

The *A Priori* Approach

The *a priori* approach involves the allotment of space resources to countries according to a pre-arranged formula via the medium of a formal treaty. This approach found its beginnings in terrestrial communications in the 1920s, but has only recently been applied, not without a struggle, to space communications.

To solve the problem of approaching chaos in European broadcasting in the 1920s, the European radiotelegraph administrations of the time met and drew up a formal plan which allotted all of the available frequencies allocated to broadcasting in Europe to the participating countries. The first such plan, the Geneva Plan, made allotments on the basis of existing stations, size of the country, population density, and economic development. When the International Telecommunication Union came into being in 1934, the European allotment plan became a part of ITU's Radio Regulations.[20]

The *a priori* approach has had only a modest additional usage in terrestrial communications, primarily confined to regional broadcasting and areas where there has been little political conflict, such as aeronautical and maritime mobile communications.[21]

The use of the *a priori* approach for satellite communications had its beginnings in the nebulous fears on the part of delegates of some of the developing countries that there would be no space resources left for their use when the time came that they could employ satellite communication. This fear, which had been articulated in an earlier UN discussion, led the 1963 ITU Extraordinary Administrative Radio Conference to pass a resolution which stated that:[22]

[All] members and associate members of the Union have an interest in and right to an equitable and rational use of frequency bands allocated for space communications [and that] the utilization and exploitation of the frequency spectrum for space communications be subject to international agreements based on principles of justice and equity permitting the use and sharing of allocated frequency bands in the mutual interest of all nations.

The fear that space resources, this time including the geostationary satellite orbit, would soon be monopolized by the richer countries under the first-come, first-served approach became an extremely controversial issue at the ITU's 1971 Space Conference. Consequently, after affirming the first-come, first-served approach for satellite communications, as mentioned earlier, the conference adopted Recommendation Spa 2-1 and Resolution Spa 2-1 which proclaimed that "all countries have equal rights in the use of both the radio frequencies allocated to various space radiocommunication services and the geostationary satellite orbit for these services" and that "registration with the ITU of frequency assignments for space radiocommunication services and their use *should not provide any permanent priority* for any individual country or groups of countries and should not create an obstacle to the establishment of space systems by other countries."[23] In addition, Recommendation SPA 2-1 requested that the next appropriate ITU radio conference address the whole issue of rights to use space resources if countries have begun to encounter "undue difficulty" in obtaining access to those resources.

Even more important to the future of the use of the *a priori* approach for satellite communication at the 1971 Space Conference were the discussions dealing specifically with the use of satellites for broadcasting. By the time of that conference it was fairly certain that it would soon be possible to broadcast television programs from a satellite directly into the homes of viewers. Extensive discussions had taken place in the United Nations and UNESCO concerning the right of a country to broadcast its programs into the homes of citizens of other countries without that other country's permission. After further discussion of the subject, and the subject of possible unintentional spillover of signals from one country to another, also considered by the many of delegations to be an unacceptable invasion of national sovereignty, the conference adopted a modification of a French and British proposal to the effect that the ITU should call a conference to plan the new broadcasting satellite service. In the minds of most delegates this was a call to negotiate an *a priori* treaty which would protect their countries from unwanted broadcasts from another country.[24]

The United States was the principal, and almost the only, country opposed to the calling of a conference for that purpose. The United States delegation argued forcefully, but unsuccessfully, that putting country "name tags" on radio frequency channels and geostationary satellite orbits would place severe restrictions on the future use of those resources.[25]

The debate was continued in the 1973 ITU Plenipotentiary Conference which designated 1977 as the year for the broadcasting satellite conference. The 1973 Plenipotentiary gave additional support to those who favored the *a priori* approach by adding to the ITU Convention a paragraph which designated radio frequencies and geostationary satellite orbits "limited natural resources" that "must be used efficiently and economically so that countries or groups of countries may have equitable access to both in conformity with the provisions of the Radio Regulations according to their needs and the technical facilities at their disposal."[26]

This, then, is the background for the first use of the *a priori* method of rights vesting for satellite communications that took place at the ITU's 1977 Broadcasting Satellite Conference.

THE BROADCASTING SATELLITE CONFERENCES

The 1977 Conference

The World Administrative Radio Conference for the Planning of the Broadcasting Satellite Service in the Frequency Bands 11.7-12.2 GHz (in regions 2 and 3) and 11.7-12.5 (in Region 1), or the 1977 Broadcasting Satellite Conference for short, began in Geneva on January 10, 1977. Almost from the very beginning it was apparent that the overwhelming majority of the 111 delegations in attendance were planning to create a formal, long-term *a priori* plan which would allot to each of the members of the ITU a portion of the 12 GHz frequency band allocated to satellite broadcasting and the appropriate geostationary satellite orbital positions.

The United States argued, as it had before, that an *a priori* plan would inhibit the development of satellite broadcasting technologies and procedures. The U.S. preferred a more flexible planning method which it felt would not only allow satellite broadcasting to develop as it should, but in so doing make enough additional space resources available for all countries that would ever be in the position to use them.

When it became apparent that the U.S. arguments were falling on deaf ears, the U.S. delegation switched tactics and began trying to convince its fellow delegates from the countries in Region 2, the Americas, to postpone drafting any allotment plan for Region 2 until a later date. The U.S. felt that additional time would allow those administrations to undertake the research necessary to produce an allotment plan that would better meet the needs of the region. With a great deal of effort, the United States, with the help of Brazil and Canada in the final week, was able to convince the other administrations of Region 2 to postpone drafting an allotment plan for Region 2.[27]

Despite Region 2's problems, the remaining delegates had no difficulty in drawing up an allotment plan for regions 1 and 3. Thirty-four geostationary satellite orbital positions, each separated by six degrees, were identified in the arc between 37 degrees West and 170 degrees East. Each country was allotted the satellite orbital positions and associated frequencies in the bands 11.7-12.2 GHz for Region 1 and 11.7-12.5 GHz for Region 3 that they felt were necessary to meet their projected needs. As would be expected, the USSR received the largest allotment.[28]

The Region 1 and 3 plans also contained numerous technical parameters dealing with link design, earth station hardware, space station hardware, and deployment, any major modifications of which must be agreed upon by all affected administrations. The two plans are intended to meet the needs of the countries involved for a period of fifteen years from the date of entry into force, January 1, 1979, and can be revised only by an ITU administrative radio conference called for that specific purpose.

While it may be doubtful whether any large portion of the allotments will be activated in the near future, or even during the lifetime of the agreement, the Region 1 and Region 3 satellite broadcasting plans are important in that they mark the first time that the right to use space resources depends not on the date of notification of a satellite system, but on whether the system conforms to a formal *a priori* plan.

The 1983 Region 2 Conference

Representatives of twenty-five administrations in Region 2 met finally in Geneva on June 13, 1983, to plan the 12 GHz band for the broadcasting satellite service in the Western Hemisphere. By July 17, the conference, known as the Region 2 Administrative Radio Conference on the Broadcasting Satellite Service, succeeded in drafting an *a priori* plan which allotted forty-eight geostationary orbital positions and 2,114 frequency channels to administrations in Region 2 in the frequency band 11.7-12.2 MHz, and established the necessary technical operating parameters.

Borrowing from information gained since the 1977 conference, and more up-to-date computer techniques, the Region 2 countries were able to produce a plan whose technical parameters were more sophisticated than those of the Regions 1 and 3 plans. The space between satellites in Region 2 were considerably narrowed, for instance, and up-links were planned in addition to down-links. As to the use of computers at the conference, the U.S. delegate writes:[29]

> The Plan could not have been developed without the extensive use of computer modeling techniques and facilities . . . . The planning experts of the U.S., Canada, and Brazil, who collaborated to produce the Plan adopted by the Conference, deserve great credit for their skill and endurance. Manning terminals in Geneva, they worked throughout entire nights utilizing the computers located in Washington and Ottawa. Over 500 "scenarios" were run on the U.S. computer. The experience of RARC-83, in this regard, should prove helpful to future international conferences.

The final plan was also more flexible than the 1977 plan. As described by Milton L. Smith III, in addition to a procedure for plan modification similar to that contained in the 1977 plan for regions 1 and 3, the Region 2 plan had three additional areas of flexibility. "First, a system which varies from the characteristics specified in the Plan, but will not adversely affect other administrations may be established. Second, a system which differs from the Plan may be established on an 'interim basis,' even though it may adversely affect the assignment of other administrations. Although agreement of the other administrations is required if increased interference could result, the procedure is simpler than that required for permanent Plan modification. Finally, some flexibility in orbital location was allowed. An administration which shares an orbital location may place its satellite anywhere within a 0.4 degree arc centered on the nominal orbital location."[30]

Two additional aspects of the 1983 conference and its results are worth noting. The first was the apparent lack of any political disputes that carried over into the technical sector, even including the often strained relations between Cuba and the United States.[31] The following statement by the head of the U.S. delegation is revealing in this respect: "No rhetoric was directed to the U.S. from any delegation. The Cubans' preferred portion of the arc was between 100° W and 115° W, but the carefully balanced Plan (worked out by computer modeling) put them at 89° W. At first they refused to move, but when asked to do so by the Chairman of the Conference, Luis Valencia of Mexico, they acceded."[32]

The second was the almost universal approval of the results by the administrations concerned, even from the United States which had so fervently opposed the idea of a priori planning in the past. As stated by the chief of the U.S. delegation:[33]

Of the eight orbital slots obtained by the U.S., five are excellent positions in the geostationary arc . . . . However, the two westernmost, at 175° W and 166° W, though capable of covering the Pacific Time Zone are somewhat too far west for satellites at those positions to cover all the Mountain Time Zone with as high an elevation angle as originally specified. (The U.S. had sought eight orbital positions from each of which its satellites could "see" one-half of the continental U.S. at an elevation angle of more than 20°.) The easternmost of the eight orbital positions, at 61.58° W, while a good position for coverage, is non-eclipse-protected and subject to blackouts in prime evening time during the Spring and Fall equinoxes. But, like many delegations at the Conference, the U.S. had to accept compromises in the interest of achieving a workable Plan.

Elaborating on the subject in more detail, the U.S. delegate had this to say:[34]

The U.S. objectives--to come home with adequate orbital/spectrum resources and procedural flexibility to meet the existing and future requirements of the broadcasting satellite service--were fully achieved at RARC-83 (Geneva). The eight orbital positions obtained--each with 500 MHz providing 32 channels--and the built-in flexibility of the regulatory procedures give U.S. operators assurance that they may go forward with confidence to provide direct broadcasting satellite service to the American public.

While the ease with which the administrations had drafted their broadcasting plan was gratifying, it did not necessarily carry over into the next area of concern, the more important fixed satellite service.[35]

THE FIXED SERVICE SATELLITE CONFERENCES

Preparatory Work

Their success in obtaining an *a priori* allotment plan for the broadcasting satellite service in Regions 1 and 3, and the promise of a conference to draw up a similar plan for Region 2, did not satisfy the countries that were worried about the future availability of space resources. The fixed satellite service soon became the next target for a change in the rights vesting procedure.

The subject became a major issue at the 1979 ITU World Administrative Radio Conference, a general conference unlike the single-issue 1977 space conference or the regional, single-issue 1983 space conference. Many of the delegates from developing countries made the same argument they had previously advanced concerning the future availability of appropriate space resources. This time, however, the argument was expanded to include other satellite services, especially the fixed satellite service; and it soon became apparent that *a priori* allotment plans were being considered.[36]

The United States, this time supported by a few other developed countries, argued as it had before that the system did not need fixing. Any rigid planning method, according to the United States, would only have the effect of inhibiting the introduction of new services and technologies that would ensure access to the appropriate space resources for all who could use them.

The outcome of the argument was a proposal by several of the developing countries, requesting that the ITU call an administrative radio conference to find a method "to guarantee in practice for all countries equitable access to the geostationary-satellite orbit and the frequency bands allocated to the space services."[37] A conference of two sessions was proposed, the first of which would:[38]

3.1 decide which space services and frequency bands should be planned;

3.2 establish the principles, technical parameters and criteria for the planning, including those for orbit and frequency assignments of the space services and frequency bands identified as per paragraph 3.1, taking into account the relevant technical aspects concerning the special geographical situation of particular countries; and provide guidelines for associated regulatory procedures;

3.3 establish guidelines for regulatory procedures in respect of services and frequency bands not covered by paragraph 3.2; [and]

3.4 consider other possible approaches that would meet the objective of resolves 1.[39]

The second session was to be called to implement the decisions taken at the first session.

In an attempt to forestall what many considered to be all but inevitable, the United States, supported by several other developed countries, read a statement at the conference stating that the words "planning" and "planned" in the resolution should be interpreted in "a broad and flexible sense" and not necessarily confined to a priori planning. In response, the delegate from India stated that, since the proposed planning conference was sovereign, it would have the right to decide which was the correct interpretation of those words.[40]

The next step was the Conference Preparatory Meeting (CPM) held in Geneva from June 25 to July 20, 1984. The 1979 WARC had requested that the ITU's CCIR, which traditionally has been dominated by representatives of developed countries, carry out preparatory studies for the 1985 Space WARC and provide the members of the ITU with the technical information that they would need to help them plan the other satellite services. Organized as a special joint meeting of interested CCIR study groups, the Conference

Preparatory Meeting was attended by delegates from sixty-one member administrations, less than half of the ITU's total membership. The final report of the Meeting, in addition to a discussion of technical and operational issues, identified twelve space services that utilized the GSO or would do so in the near future:[41]

1. fixed satellite;
2. broadcasting satellite;
3. mobile satellite;
4. radio determination satellite;
5. space operation;
6. space research;
7. earth exploration satellite;
8. meteorological satellite;
9. inter-satellite;
10. amateur;
11. radio astronomy; and
12. standard frequency and time-signal satellite.

After surveying the use being made of these services, the CPM concluded that only in the fixed satellite service was there evidence of utilization to the point where planning might be of some utility. As stated in the report:[42]

So far there is only one fixed satellite service band pair for which the utilization intensity can be regarded as relatively high, 6/4 GHz, and this intense use is largely confined to portions of the GSO providing service to certain geographical areas. The only other bands in which the intensity can be expected to approach that of 6/4 GHz is at 14/11-12 GHz, although at those frequencies the intensity is expected to rise gradually.

The CPM also identified seven possible methods for achieving the objectives laid down by the 1979 WARC. Five of the seven involved world or regional *a priori* plans valid for different time periods, and two would have relied on periodic meetings of interested administrations to work out any problems that should occur. Although there was not much discussion of the relative merits of the various systems, it should be noted that delegates from four developing countries criticized one of the two methods which did not include an *a priori* plan. The CPM also identified eleven characteristics that should be taken into account in applying the seven planning methods:[43]

1. equitable access;
2. service requirements;
3. accommodation of unforeseen new networks or changes in traffic requirements;
4. accommodation of existing networks;
5. access for multi-administration satellite networks;
6. establishment and modification of technical parameters and interference criteria;
7. restrictions due to sharing with terrestrial services;
8. restrictions due to sharing between a planned service and an unplanned space service;
9. efficient use of the orbit/spectrum;
10. impact on satellite system costs; and
11. administrative costs.

The 1985 WARC

Delegations from 110 ITU member administrations met in Geneva on August 8, 1985, for the First Session of the World Administrative Radio Conference on the Use of the Geostationary Satellite Orbit and the Planning of the Space Services Utilizing It, in order to determine the best way to assure "equitable access" to space resources for all nations.

The conference had little difficulty in establishing the technical parameters and criteria appropriate for future satellite communication systems, including those for sound broadcasting by satellites and inter-service sharing. The items discussed included: 1) space debris; 2) satellite relocation; 3) station-keeping; 4) reverse band working; 5) interference criteria; 6) polarization; 7) band segmentation; 8) frequency band pairing; 9) time-phased introduction of new technology; 10) orbit management criteria; 11) hybrid satellites; 12) spacecraft antenna performance; 13) earth station antennas; and 14) burden sharing.[44]

The conference spent a great deal of time identifying the basic principles to be used as a guide for planning, especially those which involving the "multi-administration networks" such as INTELSAT, INMARSAT, INTERSPUTNIK, and others. The final results, as summarized by the U.S. delegation, were:[45]

(1) The planning methods shall guarantee in practice for all countries equitable access to the geostationary orbit and the frequency bands allocated to the space services

utilizing it, taking into account the special needs of developing countries and the geographical situation of particular countries.

(2) The planning method and associated regulations must not impose additional constraints on terrestrial and/or space services sharing a frequency band on a co-equal basis.

(3) The planning method should consider the full-orbit/spectrum resource, and no permanent priority in the use of particular frequencies and orbital locations should be afforded to any administration so as to foreclose access by other administrations to the orbit.

(4) The planning method should take into account the relevant technical aspects of the special geographical situation of particular countries.

(5) The planning method shall take into account existing systems and, if necessary, these systems may be subjected to some adjustments to allow for the accommodation of new systems.

(6) The planning method will take into account the requirements of administrations using multi-administration systems.

(7) The planning method should provide the means to accommodate unforeseen requirements of administrations and should also be capable of accommodating advances in technology.

(8) A worldwide planning solution is preferred, but the possibility of different planning methods in different regions, frequency bands or orbital arcs is not excluded if done at a World Conference.

(9) The planning method should ensure efficient and economical use of the geostationary orbit and frequency bands allocated to space services.

(10) The planning method should be able to accommodate multi-service and/or multi-band satellite networks, without imposing undue constraints to planning.

(11) The administrative cost of the developing and application of the planning method must be as low as possible.

The conference had more difficulty in agreeing on what satellite services needed *a priori* planning. The United States argued, as it had in the past, but this time with a little more support, that no *a priori* planning was appropriate for any service. Although one delegation expressed the opinion that *a priori* planning was also indicated for the satellite mobile service, the vast majority went along with the view of the Conference Preparatory Meeting that the only service where overcrowding could be expected in the foreseeable future was the fixed satellite service. It was decided that the old first-come, first-served procedure would be continued for all other services, although, in order to assure "equitable access" for future systems, an effort would be made at the Second Session to improve the procedures involved.

The greatest difficulty occurred over the specific frequency bands allocated to the fixed satellite service that should be the subject of *a priori* planning. The United States and several other developed countries argued for a minimal approach, especially one that would not interfere with the commercial and governmental uses that were already being made of portions of the available space resources. A group of developing countries, however, proposed an *a priori* plan for all of the fixed satellite radio frequency bands below 31 GHz.[46] The final decision on this subject, a compromise, was not reached until Friday of the last week of the conference. As explained in the report of the U.S. Delegate:[47]

> This compromise was that improved coordination procedures would be applied to those portions of the 4/6 and 11-12/14 GHz bands currently being used and that an allotment plan would be established in the as yet unused expansion bands at 4/6 and 11/13 GHz. At the last minute, at the initiation of India, the 20/30 GHz commercial bands used by US Government systems were not included in the planned bands, and thus the current procedures would continue to be applied in those bands.

In particular, it was decided that the allotment plan, which "shall permit each administration to satisfy requirements for national services from at least one orbital position, within a predetermined arc and predetermined band(s)," would be established in the bands 4 500-4 800 MHz and 300 MHz to be selected in the band 6 425-7 075 MHz; and 10.70-10.95 GHz, 11.20-11.45 GHz and 12.75-13.25 GHz. The allotment plan would be for a duration of ten years.[48]

The "improved procedures" would be applied in the following bands:[49]

1) 3 700 - 4 200 MHz
2) 5 850 - 6 425 MHz
3) 10.95 - 11.20 GHz
4) 11.45 - 11.70 GHz
5) 11.70 - 12.20 GHz in Region 2
6) 12.50 - 12.75 GHz in Regions 1 and 3
7) 14.00 - 14.50 GHz

8) 18.10 - 18.30 GHz
9) 18.30 - 20.20 GHz
10) 27.00 - 30.00 GHz

The 1985 conference also directed various organs of the ITU to carry out a number of studies to assist the 1988 conference in carrying out its mandate. The IFRB was requested to develop a software package to be used in the preparation of the final allotment plan and to carry out planning exercises with that package in advance of the Second Session. The IFRB was also asked to develop guidelines for improved regulatory procedures for those portions of the radio frequency spectrum allocated to the fixed satellite service not covered by the allotment plans. The CCIR was asked to carry out a number of highly technical studies, including the development of new criteria for sharing of frequency bands by the fixed service and other services, and a study of the 20/30 GHz frequency band to help the Second Session decide whether or not it should recommend that a conference should be called to establish a plan for that frequency band.[50]

The 1988 WARC

The second of the two scheduled fixed satellite communication service world administrative radio conferences met in Geneva from August 29 to October 6, 1988, to complete the work destined to guarantee in practice to all countries "equitable access to the geostationary orbit." In all, 937 delegates from 120 countries and thirteen regional and international organizations were in attendance.[51]

Under the chairmanship of Dr. I. Stojanovic of Yugoslavia, the conference produced a Final Acts, signed on October 6, containing what the delegates felt was a good compromise plan for achieving equitable access for the fixed satellite communication service. Although the conference was much less acrimonious than the first session, it did extend one day beyond the adjournment deadline and was successful only then as a result of "brutal 18-hour sessions ending at about 3:30 a.m. during the final days."[52] The Final Acts, which become a part of the ITU Radio Regulations, will come into effect on March 16, 1990, at 0001 hours Universal Coordinated Time and are valid for a period of twenty years from that date or until revision by a competent ITU administrative radio conference.[53]

There are two components to the Final Acts of the 1988 Space WARC, an *a priori* allotment plan for fixed satellite communication services in frequency bands 4 500-4 800 MHz, 6 725-7 025 MHz, 10.7-10.95 GHz, 11.2-11.45 GHz, and 12.75-13.25 GHz, and an "improved" procedure for bringing into service satellite communication systems in certain other frequency bands.

The allotment plan is also in two parts. Part A contains 250 allotments of space resources for the 165 members of the ITU. These allotments are composed of: 1) a nominal orbital position; 2) a service area for national coverage; 3) general parameters; 4) a predetermined arc within which the definitive orbital position may be chosen; and 5) frequency bands, 800 MHz in bandwidth, required to operate the satellite.[54] The United States received two allotments under this portion of the plan, one at 159° West to provide coverage for the continental U.S. and Hawaii and Guam, and one at 101° degrees West to provide coverage of continental U.S. and Puerto Rico.[55]

Table 4.1

WARC-ORB 1988 Frequency Allotment Plan

Uplink	Downlink
6 725 to 7 025 MHz	4 500 to 4 800 MHz
12.75 to 13.25 GHz	10.7 to 10.95 GHz
	11.2 to 11.45 GHz

Part B of the plan contains a listing of "existing" fixed satellite communication systems which includes those systems 1) "for which information has been received by the ITU or those for which coordination procedure has been initiated prior to 8 August 1985 or 2) those recorded in the Master Frequency Register."[56] The United States has two such systems, one for International Satellites, Inc. (ISI) with two satellites at 56° and 58° West and one for PanAmSat with two satellites at 57° and 45° West.[57]

The overall allotment plan is accompanied by a set of procedures covering various aspects of the plan. Among these is the process for converting an allotment into an assignment. For systems in the pre-design stage in Part A of the plan, "the predetermined arc is a portion of the geostationary arc defined by the intersection of an arc segment plus or minus 10 degrees from the designated nominal orbital positions, and the allotment's corresponding service arc. When a system moves into the 'design stage,' the predetermined arc will be narrowed to plus or minus 5 degrees on either side of the nominal orbital position. For a system in the 'operational stage,' the predetermined arc will be set at zero."[58]

Two of the major conflicts at the 1988 Space Conference involved existing systems and additional uses. As regards existing systems, the problem and its solution are described by the head of the U.S. delegation as follows:[59]

> The tension between the desires of developed and developing nations was focused on the issue of existing systems. Developed nations, which comprised the majority of administrations with existing systems, wanted those systems included on a par with allotments in the Plan. Developing countries were concerned that such accommodation would result in a less than optimum Plan or in some manner give precedence to existing systems over national allotments. The debate focused in two areas--the length of operation of existing systems and the manner in which such systems would interact with allotments in the Plan. Ultimately a compromise was reached on the length of operation, led by efforts of administrations with existing systems. This compromise related the life of the Plan--a minimum of 20 years from the date of coming into force--with the life of existing systems--20 years from the date of coming into force. The other concern--that existing systems would take precedence over allotments in the Plan--was addressed through strong wording that existing systems would bear the burden of making adjustments needed to coordinate with allotments in the Plan when they are converted to assignments.

Since it was obvious that few of the members of the ITU would be in a position to utilize their allotments in the near future, the countries with existing systems, and those which expected to create systems pushed for an arrangement which would permit them to make additional use of some of the allotments of those countries which were not yet able to use them. After extensive debate, a compromise was reached whereby developed countries were permitted some use of the Allotment Plan for systems that might not be in conformity with the allotments in the plan. As a part of that compromise, the additional use systems were limited to a fifteen-year life and the additional use provisions would not be used until one year after the coming into force of the Plan. As stated by the head of the U.S. Delegation: "This moratorium addressed the concern of some administrations that developed countries would start a landrush of additional use filings which would use up any flexibility remaining in the Plan prior to implementation of allotments by developing nations."[60]

The procedure for bringing an allotment into an assignment also specifies that if the assignment is compatible with Part A of the Plan but not Part B of the Plan, "An administration responsible for an existing system or an additional use shall, depending on the stage of the development of its system, take all technically and operationally possible measures to remove incompatibilities at the pre-design, design, and operational stages in order to accommodate the requirements of the administration seeking to convert its allotment into an assignment."[61]

The plan also provides for the creation of subregional fixed satellite communication systems by groups of member administrations. Subregional in this sense involves the creation of a satellite system among neighboring countries to provide domestic or regional services within the geographical areas of the countries concerned. The procedure requires that the group "select one or more orbital positions for the system, preferably from the national allotments concerned." One of the involved administrations would be designated as the notifying administration for the group. "All or part of the national allotments used by the subregional system shall be suspended for the period of operation of this subregional system unless it can be used in a way that does not affect allotments in the plan or assignments made in accordance with the procedures associated with the plan."[62]

"Suspended national allotments . . . shall continue to enjoy the same protection as that afforded to other allotments in the Plan which are not suspended, for use in the event of cessation of the subregional system."[63]

"When administrations participating in a subregional system terminate that system and the IFRB is notified, part A of the plan would be modified to indicate that the national allotments previously designated for use by the subregional system no longer were suspended. If an administration decided, however, to withdraw from a continuing subregional system in order to implement its own national system, it would have to proceed under procedures for additional uses for the allotment."[64]

The Allotment plan also contains a procedure to permit new members of the ITU to obtain an allotment. According to this procedure, the new member submits a request to the IFRB containing the following information:[65]

1) the geographical coordinates of not more than 10 test points for determining the minimal ellipse to cover its national territory;

2) the height above the sea level of each of its test points and the rain zone or zones;
3) any special requirement, other than a fixed orbital position is to be taken into account to the extent possible.

On the basis of this information, the IFRB shall find an appropriate orbital position and enter the allotment in Part A of the Plan. In so doing, the Board shall consult, "and if necessary seek the agreement of," any administration that may be affected.[66]

The allotment plan was given a duration of twenty years, from March 16, 1990, to March 15, 2010, or until revised by a future ITU conference. Existing systems were given a beginning as of the date of filing with the IFRB; additional uses may not begin before March 16, 1991, and will end fifteen years from date of implementation.

The second component of the Final Acts of the 1988 Space WARC consists of the "improved procedures" for the remaining frequency bands which remain under the old first-come, first served system. The main element in the improved procedures was the addition to the procedures already in place in Article 11 of the Radio Regulations of the "Multilateral Planning Meeting." The holding of multilateral planning meetings, involving all administrations which feel that their existing or planned satellite systems might be affected by a new satellite system, was first advocated by the United States at the 1985 WARC as a substitute for the *a priori* allotment plan desired by many of the administrations present at that conference. As we have seen, the United States was not successful in that endeavor.

Between the 1985 and 1988 conferences a number of administrations had begun to look on multilateral planning meetings as the answer to the problem of improving the procedures for implementing new satellite systems in the frequency bands that would not be included in the forthcoming allotment plan. "Some countries appeared to be interested in something akin to a WARC, with ITU members entitled to a vote, regardless of whether they were directly affected, and with costs born by the ITU."[67] In order to counter what it considered would be an inefficient and costly process, the U.S. initiated a series of bilateral discussions in which it advocated the adoption of a more informal, less costly procedure.

The United States succeeded in its endeavor, and the 1988 Space WARC adopted an addition to the procedures already in the Radio Regulations, which simply authorizes an administration proposing a new satellite system to call a multilateral planning meeting if it encounters a "major difficulty in obtaining coordination."[68] The place of the meeting would be decided by, and the cost would be borne by, the participants. One innovation was the authorization of the participation in the meetings of multiadministration systems such as INTELSAT.

In other actions, the 1988 WARC adopted a Feeder Link Plan for the broadcasting satellite service for Regions 1 and 3 and endorsed a worldwide allocation for HDTV in a band in the range 12.7-23 GHz to be selected by a future conference. It was decided also that a conference, to be held no later than 1992, should select a band or bands of frequencies in the 500-3,000 MHz range for use by satellite sound broadcasting services.

SUMMARY CONCLUSIONS

Space resources, especially orbital positions and the appropriate radio frequency bands, have an international component by virtue of their very nature. Satellite communication operations must be kept apart if they are to function properly. The expense of developing a satellite communication system is such that no administration or private operating agency would want to see the functioning of its system jeopardized by harmful interference.

The ITU conferences which address these issues, as we have noted, are not confined to representatives of countries that either have their own satellite communication systems or will have them in the foreseeable future. The potential benefits of satellite communication have attracted almost every country in the world to the ITU satellite communication conferences to make certain that their future needs are not overlooked. The fact that each country has only one vote gives the developing country majority a substantial role to play in such conferences, ensuring that their needs and desires cannot be ignored.

So far, at least, the delegates to the ITU conferences have been able to devise procedures that seem to have met the needs of all members, including both those of the developing countries and those of the more developed countries such as the United States, without too much difficulty. Whether it can continue to do so as more countries develop their own satellite communication systems is another matter.

NOTES

1.   For an account of the early history of international efforts to regulate the use of radio see George A. Codding, Jr., *The International Telecommunication Union: An Experiment in International Cooperation*, Leiden: E.J. Brill, 1952. (Reprinted in 1972 by Arno Press, New York.)

2.   The plenipotentiary conference, meeting in principle every five years, is the ITU's supreme organ with the power to amend the ITU's basic treaty, the International Telecommunication Convention. The Administrative Council, which is made up of representatives of forty-one countries selected by the plenipotentiary conference, meets once a year and is in charge of the administration of the Union between plenipotentiaries. For additional information on the ITU's structure and functions, see George A. Codding, Jr. and Anthony M. Rutkowski, *The International Telecommunication Union in a Changing World*, Dedham MA: Artech House, Inc., 1982.

3.   In ITU terms, while "allocation" refers to the apportionment of frequencies to radio services, "allotment" refers to the frequencies provided to areas of the world or specific countries, and "assignment" to frequencies given to specific radio stations.

4.   The Radio Regulations identify thirty-six different radio services, including broadcasting, amateur, aeronautical mobile, land mobile and the like. See ITU, *Radio Regulations, Edition of 1982* (revised in 1985, 1986 and 1988), Art. 1, Section III.

5.   See Codding, *The International Telecommunication Union*, p. 95.

6.   See Codding and Rutkowski, p. 251.

7.   *Radio Regulations, Edition of 1982*, Article 8, Section IV.

8.   *Ibid.*, p. 8-142.

9.   See ITU, *Radio Regulations, Geneva, 1959*, Geneva, 1960. Footnotes for alternative uses were attached to allocations 2, 5, and 7.

10.   See U.S., Department of State, Telecommunications Division, *Report of the Chairman of the United States Delegation to the Extraordinary Administrative Radio Conference To Allocate Frequency Bands for Space Radiocommunication Purposes, Geneva, October 7, 1963 through November 8, 1963*, Washington, D.C.,December 15, 1963 (mimeo.), pp. 14-16.

11.   *Ibid.*, p. 16.

12.   See U.S., Department of State, Office of International Telecommunications Policy, *Report of the Chairman of the United States Delegation to the World Administrative Radio Conference of the International Telecommunication Union, Geneva, Switzerland, September 24--December 6, 1979*, TD Serial No. 116, Washington, D.C., 1980 (mimeo.), p. 57.

13.   From *Radio Regulations, Edition of 1982*, Art. 8.

14.   For the most recent version of this process, see *Ibid.*, Art. 12.

15.   The information about the new system that is required includes: 1) identity of the satellite network; 2) date of bringing it into use; 3) name of the administration or administrations submitting the advanced information; and 4) orbital information including planned nominal geographical longitude on the geostationary satellite orbit, the planned longitudinal tolerance and inclination excursion, and the arc of the geostationary satellite orbit over which the space station is visible, at a minimum angle of elevation of 10 degrees at the Earth's surface, from its associated earth stations or service areas. Information concerning the earth to space direction includes: 1) earth to space service area(s); 2) class of stations and nature of service; 3) frequency range; 4) power characteristics of the transmitted wave; 5) characteristics of space station receiving antenna; 6) noise temperature of the receiving space station; 7) necessary bandwidth; and 8) modulation characteristics. Information concerning the space to earth direction includes: 1) space to earth service area(s); 2) class of stations and

nature of service; 3) frequency range; 4) power characteristics of the transmission; 5) characteristics of space station transmitting antenna; 6) characteristics of receiving earth stations; 7) necessary bandwidth; and 7) modulation characteristics. See *Ibid.*, Appendix 4.

16. *Ibid.*, Art. 11, Section I.

17. *Ibid.*, Art. 11, Sections II-V.

18. Also, as in the case of a notification of a frequency to be used in a terrestrial radio service, there are provisional and appellate procedures that could result in the eventual recording in the Master Frequency Register with full or secondary rights.

19. *Ibid.*, Art. 13.

20. See Codding and Rutkowski, pp. 257-268.

21. For an account of the aborted attempt by the United States to create an *a priori* plan which would encompass almost all of the usable frequency spectrum of the time, see Codding and Rutkowski, pp. 30-35.

22. See the account given in E.D. DuCharme, R.R. Bowen, and M.J. Irwin, "The Genesis of the 1985/87 ITU World Administrative Radio Conference on the Use of the Geostationary Satellite Orbit and the Planning of Space Services Utilizing It," *Annals of Air and Space Law*, Vol. VII, 1982, p. 265.

23. ITU, *Final Acts of the Extraordinary Radio Conference for Space Communications, Geneva, 1971*, Geneva, 1981, Resolution Spa 2-1 and Recommendation Spa 2-1. Emphasis added.

24. *Ibid.*, Resolution No. Spa 2-2.

25. See U.S., Department of State, *Report of the United States Delegation to the World Administrative Radio Conference for Space Telecommunications, Geneva, Switzerland, 7 June to 17 July, 1971*, TD Serial No. 26, Washington, D.C., 1971, p. 48.

26. See ITU, *International Telecommunication Convention, Malaga-Torremolinos, 1973*, Geneva, 1974, Art. 33 (2).

27. See U.S., Department of State, *Report of the United States Delegation to the ITU Region 2 Administrative Radio Conference on the Broadcasting Satellite Service, Geneva, Switzerland, June 13-July 17, 1983*, Washington, D. C., October 31, 1983 (mimeo.), p. 7.

28. See ITU, *Final Acts of the World Administrative Radio Conference for the Planning of the Broadcasting-Satellite Service in Frequency Bands 11.7 - 12.2 GHz (in Regions 2 and 3) and 11.7 - 12.5 GHz (in Region 1), Geneva, 1977*, Geneva, 1977.

29. See *Report of the United States Delegation*, (1982) p. 3. See also pp. 55-65.

30. See Milton S. Smith, III, "Space WARC 1985 -- Legal Issues and Implications," an LL.M thesis submitted to the Faculty of Graduate Studies and Research at the Institute of Air and Space Law of McGill University, 1984, pp. 128-129.

31. The major exception involved requirements filed by Argentina which were described by test points on the Falkland Islands/Malvinas. See *Report of the U.S. Delegation* (1982), pp. 50-51.

32. *Ibid.*, p. 2.

33. *Ibid.*

34. *Ibid.*

35. Being only a regional radio conference, the results of the 1983 space WARC had to be approved by a world radio conference. This was accomplished at the 1985 Space WARC discussed below.

36. See the account in Anthony M. Rutkowski, "Ad-Hoc Two: The Third World Speaks its Mind," *Satellite Communications*, March, 1981, pp. 22-27.

37. See *Radio Regulations, Edition of 1982*, Resolution No. 3, Resolves 1., p. 1.

38. *Ibid.*, p. 2.

39. *Resolves* 1. Reads: "A world space administrative radio conference shall be convened not later than 1984 to guarantee in practice for all countries equitable access to the geostationary-orbit and the frequency bands allocated to space services."

40. See the account in Ducharme and others, pp. 274-275.

41. See ITU, CCIR, *Technical Bases for the World Administration Radio Conference on the Use of the Geostationary-Satellite Orbit and the Planning of the Space Services Utilizing It (WARC-ORB)(1)*, Report on the CCIR Conference Preparatory Meeting (CPM), Joint Meeting, Study Groups 1, 2, 4, 5, 7, 8, 9, 10, and 11, Geneva, 25 June to 20 July 1984, Geneva, 1984, Part I, pp. 9-16.

42. *Ibid.*, p. 22.

43. See *Ibid.*, pp. 36-37.

44. See U.S., Department of State, Bureau of International Communications and Information Policy, *Report of the United States Delegation to the First Session of the ITU World Administrative Radio Conference on the Planning of the Geostationary Satellite Orbit and the Space Services Utilizing It, Geneva, Switzerland, August 8 - September 16, 1985*, Washington, D.C., February 21, 1986 (mimeo.) pp. 37-40. See also ITU, World Administrative Radio Conference on the Use of the Geostationary-Satellite Orbit and the Planning of Space Services Utilizing It, First Session, Geneva, 1985, *Report to the Second Session of the Conference, Geneva, 1985*, Geneva, 1986.

45. See *Report of the United States Delegation (1985)*, pp. 53-54.

46. This included the 14/6, 17/8, 11/14, and 20/30 GHz bands. See *Ibid.*, p. 47.

47. *Ibid.*, pp. 47-48.

48. ITU, World Administrative Radio Conference on the Use of the Geostationary-Satellite Orbit and the Planning of Space Services Utilizing It, First Session, Geneva, 1985, *Addendum to Report of the Second Session of the Conference, Document 324 (Rev.1)*, 15 September 1985, p. 3.

49. *Ibid.*

50. See *Report to the Second Session of the Conference (1985)*.

51. International organizations represented included the United Nations, World Meteorological Organization (WMO), International Maritime Satellite Organization (INMARSAT), International Telecommunications Satellite Organization (INTELSAT), International Organization of Space Communications (INTERSPUTNIK), European Broadcasting Union (EBU), International Amateur Radio Union (IARU), Arab Telecommunication Union (ARU), PanAfrican Telecommunication Union (PATU), European Space Agency (ESA), Association of State Telecommunication Undertakings of the Andean Sub-Regional Agreement (ASETA), Arab Satellite Communications Organization (ARABSAT), European Telecommunications Satellite Organization (EUTELSAT), Inter-Union Commission on Frequency Allocations for Radio Astronomy and Space Science, and International Radio and Television Organization (OIRT). See ITU, Press Release, ITU/88-20, 6 October 1988, p. 1.

52. See "U.S. Pleased with WARC Results," *Broadcasting*, October 10, 1988, as quoted in *Teleclippings*, November 7, 1988, p. 2.

53. ITU Press Release, 6 October 1988, p. 5.

54. *Ibid.*, p. 4.

55. "U.S. Pleased with WARC Results," p. 3.

56. ITU Press Release, 6 October 1988, p. 6.

57. "U.S. Pleased with WARC Results," p. 3.

58. *Ibid.*

59. U.S., Department of State, *Report of the United States Delegation to the ITU World Administrative Radio Conference on the Use of the Geostationary Satellite Orbit and on the Planning of Space Services Utilizing It, Geneva, Switzerland, August 29 - October 6, 1988*, Washington, D.C., March 20, 1989, pp. 26-27.

60. *Ibid.*, p. 27.
61. *Communications Reports*, October 24, 1988, p. 21.
62. See ITU, *Final Acts Adopted by the Second Session of the World Administrative Radio Conference on the Use of the Geostationary-Satellite Orbit and the Planning of Space Resources Utilizing It (Orb-88), Geneva, 1988*, Geneva, 1988, Addendum, p. 49-50.
63. *Ibid*, p. 50.
64. *Communications Reports*, October 24, 1988.
65. *Final Acts Adopted by the Second Session (1988)*, p. 52.
66. *Ibid.*
67. "U.S. Pleased with WARC Results," p. 3.
68. *Ibid.*

CHAPTER FIVE

DOMESTIC ASSIGNMENT OF

SPACE RESOURCES

INTRODUCTION

Once the ITU has completed the allocation of portions of the available space resources to satellite communication services, it is the task of appropriate government agencies to assign those space resources to various competing domestic applicants. The ease or difficulty of this process has an important impact on the number of applicants and the services that they may be prepared to offer.

The purpose of this chapter is to describe this process as it has evolved in the United States. The total space resources the United States can make available for its own use depends on the effectiveness with which the United States makes its needs known in ITU administrative radio conferences. After introducing the major actors in the process of assigning space resources to its users, the most important being the NTIA for the governmental sector and the FCC for the private sector, a section will be devoted to the manner in which the United States participates in ITU conferences. This will be followed by sections on the procedure by which NTIA and FCC assign the space resources over which they have control. The PanAmSat story will be used to illustrate how this process works in practice.[1]

U.S. TELECOMMUNICATIONS POLICY-MAKING STRUCTURE

There are two primary actors involved in the day-to-day domestic allocation of space resources, the National Telecommunications and

Information Administration (NTIA) and the Federal Communications Commission (FCC). A third actor, the Department of State, becomes involved when it is necessary to negotiate with foreign policy makers in ITU conferences to promote the national interests of the United States.

NTIA

The NTIA, an operating unit of the Department of Commerce, derives its authority from the Communications Act of 1934, as amended, which gave the President the power to assign radio frequencies to Government radio stations. The NTIA has been given three major functions in reference to space resources: 1) making space resources assignments to the agencies of the Federal Government; 2) formulating plans and policies for the effective and efficient use of space resources in coordination with the FCC; and 3) helping develop U.S. proposals for ITU conferences. The NTIA is directed by the Assistant Secretary of Commerce for Communications and Information, who acts as the principal advisor to the President on telecommunications matters.

Within NTIA, resource matters are the primary concern of the Office of Federal Systems and Spectrum Management. The Office is headed by an Associate Administrator. Of special interest to us are the Division of Spectrum Plans and Policies, the primary instrument for preparations for ITU Conferences; the Frequency Assignment and the IRAC Support Division, which processes government requests for space resources; and the Spectrum Engineering and Analysis Division, which helps in the resolution of electromagnetic compatibility problems and prepares technical studies for national and international frequency management.

Although listed as an advisory body to the Department of Commerce, the Interdepartment Radio Advisory Committee (IRAC) is one of the more important national players. IRAC dates from 1922 and consists of representatives from twenty government departments and agencies. IRAC is charged with "assigning frequencies to U.S. Government radio stations and in developing and executing policies, programs, procedures, and technical criteria pertaining to the allocation, management, and use of the spectrum."[2] Inasmuch as the government is a major user of space resources, the IRAC plays an important role in the domestic distribution of space resources.

IRAC has three subcommittees, the Frequency Assignment Subcommittee (FAS), the Spectrum Planning Subcommittee (SPS), and the Technical Subcommittee.

FCC

The Federal Communications Commission (FCC), responsible to Congress, is an independent regulatory agency which derives it authority over civilian space resources from the Communications Act of 1934, as amended, and the Communications Satellite Act of 1962. The Commission is made up of five commissioners appointed by the President with the advice and consent of the Senate. The Commission is responsible for the management of space resources by the civilian sector and state and local governmental entities. The FCC also helps determine U.S. proposals and positions at ITU conferences.

The Commission has five operating bureaus which develop and implement specific policies and license individual stations in their particular area of interest. Of importance to satellite communication are the Mass Media Bureau and the Common Carrier Bureau. The Mass Media Bureau is responsible for developing, recommending, and administering policies and programs for the regulation of all radio and television broadcast industry services. In particular, it processes all applications for direct broadcast satellite services. The Common Carrier Bureau undertakes a similar task for "entities which furnish interstate or foreign communications services for hire," including satellite facilities. The Common Carriers Bureau also is responsible for advising the Commission on "policy and technical matters regarding the use of satellites and related facilities for both common carrier and ancillary communications services" and participation in ITU conferences.[3]

Department of State

The Department of State is responsible for determining the United States position and the conduct of United States delegations in negotiations with foreign governments and international bodies. As such, the State Department is nominally in charge of U.S. participation in ITU conferences. The major telecommunications unit in State is the Bureau of International Telecommunications and Information Policy. Created in August, 1985, the Bureau is headed by a Director with the rank of Ambassador, who reports directly to the Deputy Secretary of State.

The Role of COMSAT

A discussion of the major actors in the assignment of space resources in the United States would not be complete without a word about the role of COMSAT. COMSAT has a dual role as a private corporation responsible to its shareholders and as the representative of the U.S. government to INTELSAT and INMARSAT.[4] It is not surprising, therefore, that the two roles might come into conflict.[5]

In the beginning, for example, COMSAT used its vote in INTELSAT to oppose the authorization of separate systems and did not hesitate to attack them as serious threats to INTELSAT's future. In May, 1986, COMSAT Chairman Irving Goldstein called for strict enforcement of restrictions on the separate systems and warned that "COMSAT would not sit still" if the Reagan Administration were to relax those restrictions.[6]

In 1986, the Federal Government received a great deal of criticism for its inability to gain INTELSAT's approval of Panamsat's separate international communication satellite system.[7] This was followed in the same year by hearings in the House of Representatives where it was suggested that the Communications Satellite Act of 1962 might need some important amendments. Changes in the Act were opposed, however, by representatives of the State Department, Commerce, and the FCC, all of whom pointed out that the COMSAT "instruction process" had been greatly improved. The representative of the State Department added that State was now sending an observer to all meetings of INTELSAT's Board of Governors in addition to the regular delegation from COMSAT.[8]

OBTAINING THE U.S. SHARE

In the preceding chapter we discussed the role of the International Telecommunication Union in the allotment of space resources in certain frequency bands to specific countries for their broadcasting and fixed satellite services. The actual amount of these resources that the United States obtains, therefore, depends on the manner in which the responsible authorities present the U.S. case in the pertinent ITU conferences. The 1985 Space WARC will be used as an example of how this is done.

The two major objectives of the U.S. in any particular ITU conference dealing with space resources are 1) to obtain international recognition of and protection for new and existing U.S. radio systems, and 2) to achieve the adoption of allocations and allotments that "realistically advance our economic and security goals."[9]

Preparation

There are four primary agencies involved in the preparation of U.S. positions in ITU administrative radio conferences, the Federal

Communications Commission (FCC), the National Telecommunications and Information Agency (NTIA), the Interdepartment Radio Advisory Committee (IRAC), and the State Department. Preparations for an ITU radio conference can be initiated by any one of these agencies.

In the case of the 1985 Space WARC, for example, the first formal activity was the activation of working group Ad Hoc 178 by IRAC on March 11, 1980. Ad Hoc 178, which thereafter met on a monthly basis, "provided the forums for the articulation of technical views by the principal Government agencies having interests in the Conference . . ."[10] Ad Hoc 178's position statements on U.S. objectives and the issues likely to rise in the conference were transmitted to the U.S. delegation.

The Federal Communications Commission also began its preparations in 1980 with the issuance of a Notice of Inquiry soliciting suggestions from the private sector concerning issues likely to be raised in the Conference. In mid-1980 the FCC established an Advisory Committee, made up of "every interest in the field of communication satellites," including manufacturers, operators, users, and the academic community, "to provide even greater private sector input in support of its work for the conference."[11] The Committee issued its second and final report in January, 1985. In its proceedings, the FCC issued four Notices of Inquiry and a Report in Order containing its recommendations on U.S. positions and strategies for the conference.

The State Department's activity also began in 1980 with the formation in June of that year of the State Department Coordinating Committee for Future Radio Conferences. Consisting of representatives of State, NTIA, and FCC. The Committee was responsible for coordinating U.S. preparations for the numerous ITU radio conferences that had been scheduled by the 1979 WARC, including the 1985 Space WARC. State became more intimately involved in the preparations for 1985 in early 1983 with the creation of the Office of Coordinator for International Communications and Information Policy. This Office was responsible for the creation in late 1983 of a State Department Space WARC Working Group and in early 1984 of an Interagency Coordinating Group including representation from State, NTIA, NASA, USIA, DOD, and CIA.

The final phase of U.S. preparatory activity began with the naming of Dean Burch as the Chairman of the U.S. delegation in June, 1984. After approval by the White House, Dean Burch then formed a core group drawn from participating agencies and the private sector to provide the focal point for conference preparations in the final twelve months.[12] This group, meeting an average of once a week, was responsible for consultation with other countries to ascertain their views on the upcoming conference and for the formulation of the U.S. position to be taken on the various issues that were likely to be raised at that conference.

The actual delegation sent to Geneva to promote U.S. interests was composed as follows:[13]

1. *Chairman*: Dean Burch, former Chairman of the FCC.
2. *Eight Vice Chairmen*: three from State; two from the private sector; two from NTIA; and one from the FCC.
3. *Rank and File Delegates*: twenty from the private sector; thirteen from the FCC; eight from State, including three from the Geneva Permanent Mission; five from NTIA; three from the military; and one each from NASA, USIA, and the National Security Council.

The problems involved in creating a U.S. delegation to an ITU conference is graphically illustrated in the following excerpt from the report of the head of the U.S. delegation to the 1979 WARC:[14]

The difficult process of selecting delegates and determining the size of the delegation and support team was based on several criteria. First was the need for expertise and experience in spectrum management generally, and in the specific areas addressed by the WARC agenda and U.S. proposals. This expertise was not confined to professional engineers, though they comprised the largest group in this category. A second need was experience in general political and foreign affairs independent of specific background in ITU activities. Third was the need for representation of various domestic interests, including federal agencies, industries, and general public groups, that would be significantly affected by the WARC. Fourth, there was a mandate to seek out women and minorities. Finally, there was the Department policy to keep the accredited delegation to the minimum number necessary to accomplish the mission.

Results

Without knowing all that goes on in the deliberations of the various agencies involved in preparing for ITU administrative radio conferences, it is difficult to evaluate the exact nature of U.S. successes or failures. However, the following excerpts from the reports of U.S. delegates to two recent conferences will provide at least a general idea of some of the problems that U.S. delegations confront at ITU space resource conferences in protecting what they regard as the best interests of the United States.

The first is the 1979 World Administrative Conference, the last ITU conference to have as its task the revision of the whole of the International Frequency Allocation Table. U.S. objectives, according to the U.S. delegate, were obtained either in whole or in substantial part. The general work that was done on that Table, he said, "will provide a sound technical and regulatory framework for the expansion of communication facilities in the U.S. and abroad in the coming years while maintaining a significant degree of order in use of the spectrum allocations by the many radio services." The Delegation also found "that the frequency allocations for defense purposes were more vulnerable than others, since military systems typically occupy wide bandwidths in highly desirable parts of the spectrum and they are less understood by developing countries. Defense interests were not seriously harmed by the WARC decisions, but they were affected in a number of ways, including two that required the U.S. to submit reservations. Within the United States, satisfaction of future military requirements should be approached with a more flexible attitude with respect to compliance with the Radio Regulations, recognizing that this may entail additional economic and technological costs."[15]

These reservations, however, were not as serious as they might seem at first glance:[16]

> The United States submitted reservations on a small number of Conference decisions that could adversely affect an important national communications requirement. A reservation in this case is a formal statement that we will not be bound by a particular conference decision. However, too much should not be made of these reservations. It will be some time yet before the United States must decide if it will be necessary to follow through with the various reservations. Although the U.S. previously had taken only one reservation at a WARC, this was an unusual record we could not hope to have maintained. What is noteworthy is that of the hundreds of particular decisions, many of them significant, only a small handful were unacceptable to us.

The second conference was the ITU's 1983 Regional Broadcasting Satellite Service Conference convened to allot frequencies and satellite orbits to the countries of Region 2, the Americas. It will be recalled that the United States lost its fight against having the 1977 World Administrative Radio Conference draw up an *a priori* allotment plan and, after the conference decided to go ahead, convinced the countries of Region 2 to postpone drawing up a regional plan.

The United States came to the 1983 conference with three major requirements for the resulting allotments.

First, the U.S. wanted the 12.1-12.3 GHz band split in half in order to provide both the fixed satellite and the broadcasting satellite service with 500 MHz of contiguous spectrum.

Second, the U.S. wanted eight geostationary orbit positions, each with 500 MHz of contiguous spectrum, in the ellipse-protected portion of the arc serving North America.

Third, "Any plan resulting from the Conference should give the U.S. flexibility to implement a variety of satellite systems, including systems with different satellite powers, transmission formats, and coverage areas."[17]

At the end of the Conference, the U.S. delegation was able to report the following results:[18]

1. The Conference agreed that the band 12.1 GHz to 12.3 GHz be split at 12.2 GHz so as to provide 500 MHz in which to plan the Broadcasting Satellite Service. (It also agreed that this decision would be revisited if each administration could not be given at least 4 channels per service area in the final plan.) This decision was taken at an early point in the Conference, in no small part due to vigorous U.S. Delegation efforts.

2. The plan adopted by the Conference provided for up-link and for down-link. It included 48 orbital positions, with up to 32 channels assigned to one or more Region 2 administrations at each position. This was over twice as many channels (with less bandwidth per channel) than had been agreed at WARC-77 for Regions 1 and 3 combined.

3. In the Plan, the U.S. obtained eight orbital positions, five of them in prime locations (good elevation angle, good eclipse protection) with acceptable interference margins . . .

4. In the compromise making possible this plan, the U.S. accepted two Western positions (175° West and 166° West) with low elevation angles, and an eastern position (61.5° West) which has eclipses at 9-10 P.M. at times of the equinoxes but from which a satellite can cover the eastern and central time zones with very good elevation angles.

5.   All of the U.S. satellite positions have a minimum
of 9° spacing.   In addition, six positions allow
coverage of half the continental U.S. The net result
is that most parts of the continental U.S. may be
capable of receiving up to 128 channels.

As summarized by the General Accounting Office:[19]

The U.S. delegation at the RARC-83 conference fulfilled its
major pre-conference objectives of obtaining adequate
geostationary orbit and frequency capacity and providing for
procedural/technical flexibility and minimal coordination with
other nations to begin or change broadcasting satellite service.
The delegation was less successful, however, in achieving
conference agreement on the desired power level for the
satellites' signal and locations for three of eight orbital slots
assigned to the United States.

ALLOCATING THE U.S. SHARE

The U.S. share of space resources is allocated to government users
by the National Telecommunications And Information Administration
and to the private sector by the Federal Communications Commission.

Providing for Government Needs

NTIA considers the radio spectrum portion of space resources to
be "a limited natural resource" which is accessible to all nations.
Consequently, NTIA is mandated by the Presidential Reorganization
Plan No. 1 of 1977 and Executive Order 12046 of March 26, 1978, to:
"help attain coordinated and efficient use of the electromagnetic
spectrum and the technical compatibility of the communications
satellite system with existing communications facilities both in the
United States and abroad."[20] In addition, one of the major goals of
the Communications Satellite Act of 1962, which the NTIA is obliged
to respect, is to ensure "efficient and economical use of the
electromagnetic spectrum."[21]

Further, NTIA considers its mandate to be to develop and
administer radio frequencies in such a way as to "maintain a free
democratic society and to stimulate the healthy growth of the Nation,
while ensuring its availability to serve future requirements in the best
interest of the Nation."[22] More specifically, NTIA has eleven primary
objectives.[23]

1) to enhance the conduct of foreign affairs;
2) to serve the national security and defense;
3) to safeguard life and property;
4) to support crime prevention and law enforcement;
5) to support the national and international transportation systems;
6) to foster conservation of natural resources;
7) to provide for the national and international dissemination of educational, general, and public interest information and entertainment.
8) to make available rapid, efficient, nationwide, and worldwide radio communication services;
9) to promote scientific research, development, and exploration;
10) to stimulate social and economic progress; and
11) in summary, to improve the well being of man.

This is translated by the NTIA Manual of Regulations and Procedures for Federal Radio Frequency Management into the following priorities for assignment of radio frequencies:[24]

1. Frequencies used primarily for national defense and security.
2. Frequencies used primarily to safeguard life and property in conditions of distress.
3. Frequencies used primarily to safeguard life and property in other than conditions of distress where other means of communication are not available.
4. Frequencies used in scientific research which is considered necessary or desirable in the national interest.
5. Frequencies used for all other purposes which are judged upon the merits of the intended use.

As regards space resources in particular, NTIA has developed a fairly complicated approach to their management. First, any government agency that plans to develop a major new telecommunications system sends its proposal, including as much detailed technical and employment information as possible, to IRAC. IRAC shares this information with all of the government agencies involved and solicits their comments. The results are reviewed by IRAC's Spectrum Planning Subcommittee for conformity with international and national regulations and to see whether it can be operated without obstructing the operation of other satellite systems.[25] If the results of the review are favorable, NTIA will permit the agency in question to develop the system. If it is unfavorable, NTIA will either deny permission to proceed or will provide suggestions to make it acceptable.

This procedure is repeated as necessary at the conceptual, research and development stage.

At the operational stage, when specific frequency assignments and geostationary orbital positions are being applied for, IRAC's Frequency Assignment Subcommittee reviews the government agency's plans one more time for conformity with international and national regulations and compatibility with existing and planned satellite systems. If the finding is favorable, NTIA will make the appropriate frequency and orbital assignment. If unfavorable, NTIA will deny the application. Each frequency assignment is limited to a period of not more than five years, after which it must be reexamined by the Frequency Assignment Subcommittee to determine whether it should be continued. As would be expected, the greatest number of government applications come from the Navy, Air Force, and NASA.

All requests for space resources that have been approved by NTIA are sent to the FCC for transmission to the ITU.[26]

NTIA also monitors the use of space resources by government agencies to make certain that they are being used effectively, to determine potential interference situations, and to develop approaches for better management.

Private Sector Use

Those in the private sector who wish to develop an international satellite communications system must receive the approval of the Federal Communications Commission.

The FCC follows the basic "public convenience, interest, or necessity" criteria set forth in the Communications Act of 1934 when making allocations of space resources. The FCC has developed six general criteria to use in determining whether the public interest will be served in such a decision:[27]

1. Does the communication service in question really require radio spectrum or can practical substitutes, such as wire or cable, be used instead? If wirelines can be used, there may be less need for spectrum.

2. Is the communication service necessary for the safety of life and property, such as for use by police and

fire departments? Safety services receive more consideration than convenience or luxury services.

3. How many people will benefit from the service? FCC usually decides in favor of the service proposing to serve the most people when all other factors are equal.

4. Does the service meet a substantial need, and is it likely that the service can be established? Given the limited availability of spectrum, FCC desires that a service be publicly accepted and used.

5. Where in the radio spectrum can the communication service best operate? A radio service may be able to use frequencies in different parts of the spectrum, but the propagation characteristics of some frequencies may be more suitable than others.

6. What will be the financial costs, time involved, and other expenses involved in requiring an existing service to relocate to another part of the spectrum in order to make room for a new service? Relocating to another part of the spectrum generally requires equipment and other changes for existing licenses.

It is noted, however, that these six criteria are used only for broad policy guidelines rather than as a rigid formula. "As such, their application depends, to a large extent, on the issues involved in a particular allocation proceeding. For example, although the criteria are not ranked in any order of priority, in an allocation proceeding certain criteria can take on special importance while others may not be considered at all because they are not relevant. In addition, in some cases other national or FCC policy goals, such as national security or universal telephone service, may override the general allocation criteria."[28]

Individuals from the private sector who wish to operate a satellite communications system must supply the Federal Communication Commission with information: attesting to their legal capacity, demonstrating that their system will meet certain technical standards, and validating their financial qualifications.

To demonstrate an applicant's legal capacity it is necessary to file with their applications a Common Carrier and Satellite Radio License Qualification Report (FCC Form 430), which provides the minimum information that the FCC considers is necessary to make a determination of the applicant's legal capacity to operate a satellite system. Since international separate systems will not be common carriers, the applicant is not required to respond to the questions specifically pertinent to applications for common carrier satellite systems.[29]

Applicants must meet all of the technical standards, including such factors as choosing the appropriate orbital position and orbital spacing, choosing the appropriate radio frequencies, and meeting frequency tolerances and emissions limitations. Other factors include power limits, minimum angle of antenna elevation, power flux density limits, station identification, antenna performance standards, and maximum permissible interference power.[30]

The third factor, financial qualifications, has proved to be the most difficult for applicants desiring to enter the international satellite communication service. According to the FCC, "the applicant must demonstrate its financial preparedness to assume the costs and liabilities involved in constructing, launching and operating the system for one year. The applicant must show financing currently available for the planning, construction, launch and operation of the proposed system as well as the income or revenues anticipated from the operation of the system."[31]

In view of the uncertainties occasioned by the need to coordinate international separate systems with INTELSAT, the FCC has adopted a two-stage approach to determining the financial qualifications of the applicant.[32] The first stage involves "minimum financial qualifications." "The applicant must show: (1) the estimated costs of proposed construction and launch, and any other initial expenses for the proposed space station(s); (2) the estimated operating expenses for one year after launch of the proposed space stations(s); and (3) the source(s) or potential source(s) of funding of the proposed system for one year, which would include the identity of financiers and their letters of financial interest."[33] The applicant has forty-five days in which to supply this information.

The second stage involves the proposal to construct the system. The FCC will not issue an order permitting construction unless the applicant can demonstrate a current financial ability to meet the costs of construction, launch, and operation for one year. This includes: "the submission of a balance sheet, verified by affidavit, [which] shall demonstrate its current financial ability current for the latest fiscal year and documentation of any financial commitment reflected in the balance together with an exhibit demonstrating that the applicant has current assets and operating income sufficient to meet its estimated construction, launch and operation costs and expenses."[34] The applicant has sixty days following receipt by the Commission of the State Department's letter stating that the U.S. has fulfilled its obligations under the charter of INTELSAT to meet these requirements.[35] If all is in order, the Commission will issue an order permitting construction. If the applicant does not fulfill all of the conditions, the construction permit becomes null and void and the orbital positions and frequencies provisionally assigned to the applicant will become available to others.

Approval by INTELSAT and Foreign Partners

As mentioned in Chapter Two, the Charter of INTELSAT to which the U.S. is a signatory, requires that all commercial international communication satellite systems must be technically compatible with the INTELSAT system and must not cause it any serious economic harm. The negotiations on these matters are carried out by the State Department, at the request of the FCC, and involves three stages. First INTELSAT's Technical Committee reviews the application from the point of view of its technical parameters and makes a recommendation to INTELSAT's Board of Governors. If the system is for domestic service, earth resource survey, DBS or other specialized services such as RDSS, then only technical compatibility is considered and the Board of Governors takes final action. If it is for the international service, however, the Board then meets to consider both the technical and financial impact of the proposed new system. If the Board gives its approval, a meeting of the Assembly of Parties is called to decide whether or not the new system will be approved.[36]

Since satellite communication is a two-way proposition, it is also necessary for U.S. entities wishing to operate such a commercial system to find foreign partners willing to divert some of their traffic onto the new system. In most cases the partner will be the government PTT.

The problems that this poses are outlined in the following excerpt from an article by Henry M. Rivera, a former FCC Commissioner:[37]

One of the biggest challenges for separate system has been obtaining approval from foreign PTTs. Generally, the PTTs fear that additional competition will reduce their revenue and thereby make it more difficult for them to subsidize their domestic postal and telephone service. They think that dealing with separate systems will increase their transaction costs without providing any significant benefits. Many PTTs believe that the volume of international traffic will remain constant regardless of the number of U.S. carriers providing service. In some cases nationalism is also a factor--the foreign governments are reluctant to give U.S. companies a stronger foothold in providing communication service.

The requirement of obtaining the approval of a foreign government has a serious impact on the ability of an applicant to meet the FCC's financial requirements. Without the agreement of a foreign country to permit communications it is difficult to sign up customers for the proposed system, and without customers it is hard to obtain the necessary financial backing that would meet the FCC requirements.

THE CASE OF PANAMSAT

Although the procedures for the implementation of separate international satellite communication systems have been in place for a number of years, PanAmSat is the only one that is so far operational. The story of PanAmSat will therefore be used to demonstrate how those procedures operate in practice.

The PanAmSat story begins on February 13, 1984, when the Pan American Satellite Corporation filed an application with the FCC to construct, launch, and operate a satellite system providing international services. The PanAmsat was preceded by four such applications: Orion Satellite Corporation on March 11, 1983; International Satellite, Inc. on August 12, 1983; RCA American Communications, Inc. on February 13, 1984 (subsequently withdrawn); and Cygnus Satellite Corporation on March 7, 1984.[38]

PanAmSat's proposed system would have consisted of one operational satellite and one in-orbit spare. Of the thirty-six transponders, twelve would be used for communications between North and South America and twenty-four would be used for domestic service in South America. "The twelve international transponders would have 72 MHz of usable bandwidth per transponder and would be used to provide links between New York, Miami, the South American continent, and parts of Central America, the Caribbean, and the Iberian peninsula. The system would uplink at 6.4-6.9 GHz and downlink at 10.7-11.2 GHz. The satellite would be designed to provide video and audio distribution services, specifically, distribution of television and radio programs from entities such as television networks, motion picture studios, cable systems, and news and wire services. PanAmSat proposes in its application to offer its transponder capacity on a non-common carrier basis for sale or long-term lease to both U.S. and foreign customers."[39]

On April 6, 1983, shortly after the Orion filing, and on August 25, 1983, after the International Satellite Inc. filing, the Departments of State and Commerce asked the FCC to refrain from taking any action on the applications until the executive branch of the government could study their possible impact on the national interest and on U.S. foreign policy.

As mentioned in Chapter Three, the big breakthrough came in November 28, 1984, when President Reagan announced that alternative satellite systems were required in the national interest within the meaning of Sections 102(d) and 201(a) of the 1962 Communications Satellite Act. After receiving a letter from State and Commerce informing it of the President's decision and their decision that such systems should not be connected to the public switched network, the FCC issued a Report and Order in which it conditionally granted the PanAmSat's application and those of ISI and RCA.[40]

With this approval, PanAmSat accelerated its preparations for starting its system, including contracting for the modification of an existing American satellite, contracting with European Space Agency to launch its satellite, requesting American authorities to ask for approval from INTELSAT, raising the necessary finances to satisfy the FCC requirements, and attempting to find foreign countries that would permit PanAmSat to provide communication links.

PanAmSat was slow in obtaining both the necessary financing and the agreement of foreign countries to permit it to operate on their territories. In January 1987 Henry M. Rivera reported that the FCC had granted PanAmSat four extensions of its deadline to show financial support and "three and a half years after the first application was filed, it is still not certain that the separate systems will ever be able to raise enough money to begin operations."[41] And in September 1987 it was reported that PanAmSat owner Rene Anselmo had already spent some $30 million of his own money on start up operations.[42]

Concerning agreements with foreign PTT's, Rivera pointed out that "between May 31, 1984, when PanAmSat filed its application, and October, 1986, PanAmSat reached agreement with just one foreign authority: Peru. PanAmSat's agreement provides that five of the system's 24 transponders will be used for domestic communications in Peru and some communication between the United States and Peru. Even that agreement was not easy: PanAmSat ultimately had to sell at least one of its transponders to Peru for just $1 *and* PanAmSat still has 19 vacant transponders."[43]

 Obtaining the approval of INTELSAT turned out to be just as difficult. INTELSAT will approve separate satellite systems only if they are found to be technically compatible with INTELSAT and will not cause "significant economic harm" to INTELSAT. As mentioned earlier, there are three steps in this process: 1) technical and engineering specifications are submitted to INTELSAT's Technical Committee for analysis; 2) the Board of Governors meets to consider the potential technical and economic impact of the proposed satellite; and 3) the members of INTELSAT decide, in a meeting of

INTELSAT's Assembly of Parties, whether or not to approve the proposed system.

The first problem occurred when INTELSAT insisted that PanAmSat provide it with both technical and financial information for all of its proposed twenty-four transponders despite the fact that it had made arrangements to use only five. The second was a refusal by INTELSAT's Board of Governors, 66% to 20%, of an American request to call an Extraordinary meeting of the Assembly of Parties to consider PanAmSat's application.[44]

After State, Commerce, and the FCC improved their "instructional process" with COMSAT, PanAmSat's fortunes took a turn for the better. In April, 1987, INTELSAT's Assembly of Parties approved the use of five C band transponders for service between the U.S. and Peru.

The summer of 1988 saw the launch of PanAmSat's satellite. Launched on June 15 by an Ariane 4 booster from the French Space Center in Guyana, the satellite as finally constructed was different from that described in PanAmSat's original application. PAS-1, as it is known, is a hybrid, C/Ku Band satellite. "There are 18 C band transponders providing service to Latin America and the Caribbean, 6 of which are 72 MHz, and 12 of which are 36 MHz. The 6 Ku band transponders, which are divided so that three serve the Continental U.S. and three serve Europe, are all 72 MHz. There is no cross-strapping between C band and Ku band. The satellite is used for television and other program services, data broadcasting, and the two-way transmission of data (including digitized video) and voice."[45]

June, 1988, also saw a favorable recommendation from INTELSAT's Board of Governors on technical and economic coordination with PanAmSAt for six Ku band transponders for service with West Germany and the United Kingdom and five C band transponders for service between the United States and the Dominican Republic. The Board of Governors' decision was ratified by a meeting of INTELSAT's Assembly of Parties in November.

As of the time of writing, PanAmSat had completed, or was in the process of completing, INTELSAT Article XIV (d), consultations for service between the U.S. and France (for receive video service only), Ireland, Luxembourg, Sweden, West Germany, and the United Kingdom in Europe, and Brazil, Chile, Costa Rica, Dominican Republic Guatemala, Honduras, and Peru in Latin America. Article 14 (c) consultations were also under way for service in Chile, Guatemala, Honduras, and the United Kingdom.[46]

After five years of struggle, PanAmSat seems to have surmounted two of the most difficult hurdles in its quest to become the first separate international satellite communications system, gaining INTELSAT's approval and finding foreign governments which will permit PanAmSat to do business within their territories. The task of finding entities in those countries with which to do business, however, is still an important and difficult one for PanAmSat in part because of its technical limitations with regard to connectivity and the lack of sparing.[47] The more that PanAmSat can accomplish in this respect in the near future the more likely that it will be able to withstand the competition from other separate international satellite communication system which are bound to arrive on the scene now that PanAmSat has shown the way.

SUMMARY CONCLUSIONS

If U.S. satellite communications is to reach its full potential, satellite-service providers must have access to the needed space resources, orbital slots, and radio frequencies. The availability of those resources depends to a large degree on the decisions that are made in policy-making bodies on the domestic and the international levels. On the domestic level, the pertinent bodies are the FCC and NTIA. The FCC authorizes civilian requirements, while government requirements are processed through the auspices of the IRAC's Interdepartment Radio Advisory Committee. In view of the fact that government requirements usually take precedence over civilian requirements, if any government agency wishes to increase its share of the limited space resources available to the United States, all it needs to do is to convince the other government agencies in IRAC of the merits of its requests. If this is accomplished satisfactorily, the allocation will be certain to take place.

Private entities that wish to enter the international satellite communication business have a much more difficult time. They must vie for the limited amount of space resources left to be distributed by the FCC by convincing the FCC that what they are doing is in the national interest, then they must convince INTELSAT that they will not cause any technical or economic harm to it, and after that they must find foreign governments which will permit them to compete against their own systems.

On the international level the major policy-making body is the International Telecommunication Union, especially the ITU's administrative radio conferences. It is at this level where the first cut is made. The allocation of space resources to radio communications, of which satellite communication is only one service in competition with many others, and the needs of the U.S. government and its agencies are in competition with those of many other countries. Any increase in the allocation of space resources to satellite communication in this situation will occur only if those interested in bringing this about can first convince those who are responsible for the formation of U.S. policy to those conferences that such an action is desirable and also that the U.S. delegation to the conference itself can muster the necessary support among the delegations from other countries that such a policy is a valid one. Consequently, if any agency, government or private, desires to influence such an outcome, it must become an active participant in the process whereby the U.S. policy is formulated. In addition to making its views known, the agency in question should try to send with the U.S. delegation to ITU conferences dealing with space resources one or more articulate and persuasive representatives who can keep the needs of the agency in view of the delegation as a whole and who can help the delegation make a convincing case to the world at large.

NOTES

1.   The official name is Pan American Satellite. It has become customary, however, to use the abbreviation PanAmSat.
2.   U.S., Department of Commerce, National Telecommunications and Information Administration, *Manual of Regulations and Procedures for Federal Radio Frequency Management, May 1986 Edition*, Washington, D.C., 1986, 1.4.2., Art. II. The agencies involved include Agriculture, Air Force, Army, Coast Guard, Commerce, Department of Energy, Federal Aviation Administration, Federal Emergency Management Agency, General Services Administration, Health and Human Services, Interior, Justice, NASA, National Science Foundation, Navy, State, Treasury, U.S. Information Agency, U.S. Postal service, and the Veterans Administration. *Ibid.*, Art, III.
3.   U.S., Federal Communications Commission, *Rules and Regulations, Volume l*, September 1987, Sections 0.61 and 0.91.
4.   In addition to the fact that the President appoints three of COMSAT's fifteen directors, the administration also instructs COMSAT how to vote on issues that arise in INTELSAT meetings.
5.   As described by Rob Stoddard, "Responsible to its shareholders, it must also represent the U.S. government as the official signatory of INTELSAT and INMARSAT. That government now is creating competition to the global system that COMSAT created and helps to administer; the entrepreneurs that hope to compete with INTELSAT and INMARSAT allege that a conflict exists, that COMSAT can no longer serve so many different masters." Rob Stoddard, "Rethinking International Satellite Regs," *Satellite Communications*, Sept. 1987, p. 37.
6.   See Rivera, p. 42.
7.   See below for more on the PanAmSat story.
8.   Stoddard, p. 38.
9.   U.S. Senate, Committee on Commerce, Science and Transportation, *Long-Range Goals in International Telecommunications and Information*, 98th Congress, 1st Session, S. Prt. 98-22, U. S. Government Printing Office, Washington, D.C., 1983, p. 101.

10. *Report of the United States Delegation (1985), P. 12.*

11. *Ibid.*, pp. 12-13.

12. *Ibid.*

13. *Ibid.*, Annex A.

14. See U.S., Department of State, Office of International Communications Policy, *Report of the Chairman of the United States Delegation to the World Administrative Radio Conference of the International Telecommunication Union, Geneva, Switzerland, September 24 -- December 6, 1979*, TD Serial No. 116, Washington D.C., 1980 (mimeo), p. 7.

15. *Report of the Chairman of the United States Delegation to the World Administrative Radio Conference, 1979*, pp. 105-106.

16. *Ibid.*

17. In particular, the coverage areas were to include both quarter-CONUS [Continental U.S. except Alaska] and half-CONUS service from single orbital positions. *Report of the United States Delegation (1983).*

18. *Ibid.*, pp. 40-41.

19. See U.S., General Accounting Office, *U.S. Objectives Generally Achieved At Broadcasting Satellite International Conference--Improvements Can Help In Future Conferences*, GAO/RECD-84-157, Washington D.C., August 2, 1984, p.8.

20. See NTIA, *Manual of Regulations and Procedures for Federal Radio Frequency Management, May 1986 Edition*, para. 2.2.

21. *Communications Satellite Act of 1962*, Section 201, b.

22. *Ibid.*, para. 2.1.

23. *Ibid.*

24. *NTIA Manual of Regulations and Procedures for Federal Radio Frequency Management, Revisions for May 1987*, para. 2.3.6.

25. Most of this material and the material that follows on NTIA's space resource management is from the article by Buss and Cutts on radio frequency management adapted to apply to both aspects of space resources and information provided by NTIA by telephone on March 9, 1989.

26. As a result of an agreement with the State Department, the FCC relays requests for space resources directly to the ITU without going through the State Department.

27. See U.S., General Accounting Office, Resources, Community, and Economic Development Division, *Federal Communications Commission Spectrum Management*, Washington, D.C., January 12, 1989, pp. 13-14.

28. *Ibid.*, pp. 14-15.

29. FCC. *Report and Order in the Matter of Establishment of Satellite Systems Providing International Communications, CC Docket No. 84-1299, Adopted July 25, 1985*, para. 232.

30. Federal Communications Commission, *Rules and Regulations*, Vol. V, November 1987, Part 25 Satellite Communications, Subpart C.

31. *Ibid.*, paras. 233 & 234.

32. There is only one stage for determining the financial qualifications for domestic satellite applications in view of fact that no economic coordination with INTELSAT is necessary.

33. *Ibid.*, para. 235.

34. *Ibid.* There are certain modifications to these requirements for newly established entities and where the applicant is owned by more than one corporate parent.

35. These procedures of course apply only to the satellite portion of a communications satellite system, the portion we are interested in here. It is naturally much easier for U.S. applicants to apply for an earth station. For the procedure involved, see FCC, "Common Carrier Services Information: Application Filing Requirements for Domestic Satellite Earth Station Authorizations (Revised), *Public Notice, Released: June 18, 1987*, DA 87-732, Washington, D.C., 1987.

36. See the discussion of the process in Rivera, p. 40.

37. *Ibid.*

38. PanAmSat's application was also followed by one from Financial Satellite Corporation on May 17, 1985. See FCC, *Report and Order In the Matter of Establishment of Satellite Systems Providing International Communications, July 25, 1985*, CC Docket No. 84-1299, FCC 85-399.

39. *Ibid.*, pp. 6-7.

40. The FCC also had ruled that the separate systems would not be limited to intra-corporate services but would be free to offer "private-line services which are tailored or 'customized' to meet special customer needs." Rivera, p. 41. In April, 1986, the FCC expanded the range of services that separate systems were permitted to offer to include occasional-use video. This was considered a major victory for the applicants who felt that occasional-use video would be their most lucrative market.

41. Rivera, p. 41.

42. Stoddard, p. 39.

43. Rivera, p. 41. Rivera also pointed out that: "Other system applicants have fared even worse -- they have been unable to reach a single agreement with a foreign government." *Ibid.*

44. *Ibid.*

45. As set forth in a letter from Douglas Goldschmidt of Alpha Lyracom dated March 5, 1989.

46. *Ibid.*

47. In this respect PanAmSat must feel somewhat like AT&T competitors seeking to convince customers that they have superior systems.

CONCLUSIONS

As a result of its special characteristics--distance insensitivity, broadcast capability, and flexible routing--communication via satellite quickly became a primary element in the world's telecommunication network. Soon after satellites proved their feasibility, they replaced the older undersea cables as the preferred means for voice communication on long-distance high-density routes and as an exciting new method for transmitting television programs. Numerous additional applications were soon to follow: mobile satellite systems, radio determination satellite services, private data networks, domestic satellite services, and broadcasting satellite services. For a while it seemed as though satellites would monopolize long-distance communication.

However, just as satellite communication began to replace the older methods of long-distance communication, it became the subject of competition. The newer technology of fiber optic cable had its own special characteristics, including the absence of a perceptible time-lag, that made it preferable to satellite communication for a number of purposes. As a result it has begun to make inroads on the profitable Europe-North America route.

Both satellites and fiber optic cable have been the subject of important innovations. More sophisticated satellites have been developed, new services have been offered, and advances have been made in echo suppression and transmission delay. The capacity of fiber optic cable has been greatly expanded, the number of repeaters has been reduced, and techniques have been advanced that will protect fiber optic cables from being inadvertently severed.

The exact role that each technology will play in the long-term telecommunications picture depends to a great extent on how their respective technologies develop in the years to come. At present, it seems that both have important, complementary roles to play. Fiber optic cable seems best suited to high-density routes throughout the world. Satellites will continue to provide some service on long-distance, high-density routes as long as the demand for this service continues to increase. Satellites will be the preferred means for providing service for long-distance, low-density communications in both the developed and developing world and for mobile communications. Satellites will be important also in the transmission of occasional television programs and as back-up for fiber optic cable.

The ultimate resolution of the competition between the two systems and the exact role that satellite communications will play will not be dependent solely upon technology. Policy considerations, both international and domestic, will also have an essential role to play. Telecommunications are too important to the defense, the economies, and the foreign relations of states to escape from these considerations.

For a number of years the United States was in the enviable position of being able to dictate the manner in which satellite communications would develop. In the early years of satellite communications, when the United States decided it had to overcome the advantages that the U.S.S.R. had achieved with the launching of Sputnik, U.S. policy consisted of putting massive amounts of money into research and development. Later U.S. policy measures included establishing loading requirements for cable and satellites on the heavily used North America to Europe communications routes in order to encourage the use of satellites.

The creation of INTELSAT was a direct policy decision on the part of the United States. INTELSAT was given a mandate by the U.S. to seek a virtually universal membership and to exercise a global monopoly on long-distance satellite communications. INTELSAT was a gift from the United States, having the necessary wealth and technology, to the rest of the world. Despite the fact that the U.S. was to retain control of the organization in its formative years, it was still hoped that the U.S. gesture would have a favorable influence on world public opinion.

INTELSAT has achieved an impressive record over the years it has been in existence. The original fifteen countries that signed the Interim Agreement in 1964 grew to eighty-four by the time of negotiation of the Definitive Agreement, and to 117 at present. The number of channels in service has risen from 150 to 120,000 in 1989, and the cost of the annual space segment utilization charge has been reduced from $32,000 to less than $5,000. Not only has INTELSAT gained nearly half of the traffic between North America and Europe and about two thirds of the global total, but also it provides domestic satellite systems to some thirty countries. In addition, INTELSAT has been the force for new developments in satellite technology and in the creation of new services for developed and developing countries alike.

Even while the U.S. policy of making INTELSAT into a global

monopoly was still in place, however, events were occurring that made it clear that INTELSAT would never achieve a totally monopolistic position. The U.S.S.R., the first nation to put a communication satellite in orbit, failed to join INTELSAT and soon created a rival organization, INTERSPUTNIK, to meet the satellite communication needs of the Communist Bloc nations. Although INTERSPUTNIK has grown only from eight to fourteen members over the years, the mere fact of its creation spelled the end to any hope of a global monopoly for INTELSAT.

INMARSAT, which followed INTERSPUTNIK, was created to provide for the communication needs of ships at sea. A new organization was created to engage in maritime satellite communications for various reasons. Some of the founders felt that it would be necessary to include important maritime countries that were not members of INTELSAT, especially the U.S.S.R., while others thought that those countries with large shipping interests should have more voting power concerning maritime communications than they could have under INTELSAT. INMARSAT has subsequently enlarged its mandate to include aeronautical and land mobile communications. Although INMARSAT does not compete directly with INTELSAT, its creation did effectively bar INTELSAT from what could be a lucrative market for satellite communications in the future.

The regional satellite communication organization EUTELSAT was created when the European countries involved believed they had developed the necessary technology, and the Arab states were able to purchase their own satellite system as a result of the benefits they received from the petroleum price hikes in the early 1970s. Elements of prestige were also involved in the creation of both of these systems and Indonesia's Palapas enterprise.

Although it looks as though all of the regional satellite systems will probably remain restricted to regional communications, their appearance did add certain limits to what INTELSAT would be able to do in the future. And it is important to note that each of these regional systems and INMARSAT were accepted, *albeit* reluctantly, by INTELSAT.

Although each of the new systems, global or regional, diminishes to some extent the overall stature of INTELSAT, INTELSAT still remains one of the major players in long-distance communication; its activities have an important impact on the development of satellite communications and will continue to do so for the foreseeable future.

How well INTELSAT could have done with the continued full backing of its chief sponsor is a moot question, because before long the U.S. decided to reverse its earlier policy and opt for one of competition. The decision of the United States to adopt this new direction in the field of satellite communications resulted from several considerations. In the first place the United States saw its position of influence in INTELSAT diminishing as the internationally managed and staffed permanent organization became a reality. Second, U.S. presidents with a universalist viewpoint were replaced by those who favored private enterprise and the philosophy of competition. Third, there was the hope that if U.S. private enterprise were set free, it might be able to achieve a position of dominance in the field of satellite communication.

It took some time after the idea was first broached for the idea of independent domestic satellite communications systems to become a reality. However, by 1972 the pressures from various sources, including numerous individuals in the Nixon Administration, were such that the Federal Communication Commission overcame all opposition and ruled that satellites could provide a variety of existing and new services more efficiently and economically than the existing terrestrial facilities. With the help of a number of additional decisions making the start-up of a satellite system less cumbersome, the U.S. soon had a number of applications to provide satellite services. In 1989 there were thirty satellites operated by seven different companies providing a variety of domestic communication satellite services.

Pressures to allow American entrepreneurs also to provide international satellite communication systems were soon to follow, and the first application for a private U.S. communications satellite was made in March, 1983. The concept fell on sympathetic ears in the Reagan Administration, which in November, 1984, decided that separate U.S. international satellite systems were indeed in the national interest as long as they were restricted to non-switched services and bound by various other provisions designed to protect the interests of INTELSAT. The FCC followed suit and in a Report and Order in 1985 and a Memorandum and Order in the same year made separate international satellite systems official U.S. policy.

The implementation of the policy ran up against two major barriers. The first was INTELSAT itself. The member nations of INTELSAT and its signatories were reluctant to condone competition to INTELSAT, even competition restricted to non-switched services. The second barrier was a hesitation on the part of foreign telecommunication administrations to permit the new American satellite service entities to provide communication services to and from their territories.

Little by little, after a great deal of effort on the part of the United States, these barriers were successfully overcome. In 1988 the PanAmSat system was approved by INTELSAT and, at the time of writing, the first PanAmSat satellite was in orbit and PanAmSat seemed to have convinced an impressive number of countries to permit it to provide long-distance communications for them. Whether other American companies will be given the same rights is another matter. Another consideration is that foreign satellite communication entities might well come into existence and demand reciprocal rights from the U.S. If this should occur, then there would be real competition in satellite communication services.

Policy decisions of a single country--no matter how large or how far advanced in technology--or a group of developed countries, or an international organization such as INTELSAT, cannot alone decide the manner in which satellite communications will develop. There is also a heavy dependence on the availability of the necessary space resources, especially the proper radio frequencies and satellite orbital positions, which brings us to the International Telecommunication Union. The ITU is the instrumentality that has assumed the task of developing and administering the procedures involved in the allocation of space resources to the countries that require them and the prevention of harmful interference between the services that utilize them. These procedures are formulated and revised at the ITU's periodic world and regional administrative satellite communication conferences. Admission to these conferences is open to all of the members of the ITU, and each participant country has only one vote, without regard to its size or wealth.

In the beginning, the older terrestrial radio method of giving preference to the requests for space resources that were first in time was applied to satellite communications. It was not long, however, before countries without their own satellite systems began to worry that the big users of satellite communications, such as the United States, the U.S.S.R., and INTELSAT, would have monopolized all of the preferable space resources by the time they could afford them, making it difficult for them to introduce their own systems. The result was the development by the ITU of the formal *a priori* method of allotting specific portions of the radio frequency spectrum and satellite orbital positions to specific countries. The *a priori* allotment plan adopted for broadcasting satellite systems in 1977 was so rigid, however, that it aroused the apprehensions of the countries with existing satellite systems and those who expected to have them in the foreseeable future. As a result, the *a priori* plans adopted in 1983 Region 2 broadcasting satellite conference and in the 1988 fixed service satellite conferences were made more flexible, giving better protection to existing systems and even permitting uses which do not conform to a plan under certain conditions. The future of satellite communications will continue to depend on how well the nations of the world, both the developed and the lesser developed, continue to balance the needs of satellite communications with other types of communication in conferences under the auspices of the ITU.

The configuration that the future satellite communication network will take is dependent on more than decisions at the international level. In the last analysis, domestic communications policy decides just who will be permitted to use satellites for communications and just how they will be employed. Individual government policies shape the rules and regulations from the most elementary to the most complex in almost every aspect of the satellite business. Tariffs, service conditions, and the licensing of earth stations are all tightly controlled by the appropriate government agencies. Domestic authorities also decide on what organs in the state will have access to the space resources that the state has the power to allocate.

The United States has a unique structure in this respect. The U.S. agencies involved in policy making for satellite communication include the Department of Commerce, the Federal Communication Commission, the State Department, and COMSAT. All of these agencies have their own agendas which sometimes may appear to be in conflict. The Commerce Department has the dual role of promoting U.S. trade, including private satellite services, while at the same time providing for and protecting government satellite communications. The Commerce Department also decides on the amount of the allocated space resources that government agencies, including the military, are to receive. The Federal Communication Commission is responsible for the use of satellite communications in the best interest of the United States, obtaining the maximum portion of the radio spectrum and orbital positions it can for the private sector and allocating what it can obtain to the various private entities that desire them. The State Department is concerned with basic foreign policy issues and how they relate to satellite communications and U.S. activities in the International Telecommunication Union. Lastly, COMSAT is at one and the same time a commercial enterprise interested in making the maximum profit for its stockholders and the representative of the U.S. in INTELSAT and INMARSAT. It is no wonder that at times it is difficult to tell just who is responsible for making U.S. satellite communications policy and defending it in the international arena.

In conclusion, it is apparent that the development of the appropriate technology is thus only one factor involved in assessing the manner in which satellite communication will evolve in the future. A number of policy factors will also be involved. Much will depend on how the U.S. government views the value of satellite communication to government agencies, the economy, and overall foreign policy. As the world's leading producer and consumer of telecommunications equipment and services, the policy decisions the U.S. government makes will always have an important impact in this respect.

The future of satellite communications will be determined by the advances in technology that are made by countries other than the U.S. and the direction that their policy decisions subsequently take, especially in the realm of the promotion of domestic and international competition. Of major significance will be what INTELSAT decides is in its best interests and how the nations of the world cooperate in allocation of space resources at ITU conferences and meetings. The future will also be influenced by the manner in which individual nations allocate their share of those resources to potential domestic users.

Of great importance will be additional factors not discussed here. The success or failure of the efforts of European states to achieve certain elements of unity in 1992 and beyond, the economic and political developments in the Pacific region, and the changes that will occur in the U.S.S.R. and Eastern Europe will all affect the communications of the entire world. The evolution of the economic situation of the lesser developed countries will also have a role to play.

One thing is certain, and that is that in virtually every sense the future of satellite communications will be an exciting one to observe and experience.

APPENDIX 1

FREQUENCIES ASSIGNED TO SATELLITE COMMUNICATION[1]

2 501-2 502 kHz:	Space Research, secondary, shared.
5 003-5 005 kHz:	Space Research, secondary, shared.
7 000-7 100 kHz:	Amateur Satellite, primary, shared.
10 003-10 005 kHz:	Space Research, secondary, shared.
14 000-14 250 kHz:	Amateur Satellite, primary, shared.
15 005-15 010 kHz:	Space Research, secondary, shared.
18 052-18 068 kHz:	Space Research, secondary, shared.
18 068-18 168 kHz:	Amateur Satellite, primary, shared.
19 990-19 995 kHz:	Space Research, secondary, shared.
21 000-21 450 kHz:	Amateur Satellite, primary, shared.
24 890-24 990 kHz:	Amateur Satellite, primary, shared.
25 005-25 010 kHz:	Space Research, secondary, shared.

---

[1] See *Radio Regulations, Edition of 1982* (Revised in 1985, 1986, and 1988), Art. 8. In this Appendix, Radio Navigation Satellite Services are a component of Radio Determination Satellite Services referred to earlier in the text.

28-29.7 MHz:	Amateur Satellite, primary, shared.
30.005-30.01 MHz:	Space Operation (satellite identification), primary, and Space Research, primary, shared.
37.5-38.25 MHz:	Radio Astronomy, secondary, shared.
39.986-40.02 MHz:	Space Research, secondary, shared.
40.98-41 015 MHz:	Space Research, secondary, shared.
73-74.6 MHz:	Radio Astronomy, exclusive.
137-138 MHz:	Space Operation (space-to-Earth), primary, Meteorological Satellite (space-to-Earth), primary, Space Research (space-to-Earth), primary, shared.
138-143.6 MHz:	Regions 2 and 3, Space Research (space-to-Earth), secondary, shared.
143.6-143.65 MHz:	Space Research (space-to-Earth), primary, shared differently in regions.
143.65-144 MHz:	Regions 2 and 3, Space Research (space to Earth), secondary, shared differently in regions.
144-146 MHz:	Amateur Satellite, primary, shared.
149.9-150.05 MHz:	Radionavigation Satellite, exclusive.
150.05-153 MHz:	Region 1, Radio Astronomy, primary, shared.
267-272 MHz:	Space Operation (space-to-Earth), secondary, shared.
272-273 MHz:	Space Operation (space-to-Earth), primary, shared.
322-328.6 MHz:	Radio Astronomy, primary, shared.
399.9-400.05 MHz:	Radionavigation Satellite, exclusive.

400.05-400.15 MHz:	Standard Frequency and Time Signal Satellite, (400.1 MHz), exclusive.
400.15-401 MHz:	Meteorological Satellite (space-to-Earth), primary, Space Research (space-to-Earth), primary, and Space Operation (space-to-Earth), secondary, shared.
401-402 MHz:	Space Operation (space-to-Earth), primary, Earth Exploration Satellite (Earth-to-space), secondary, and Meteorological Satellite (Earth-to-space), secondary, shared.
402-403 MHz:	Earth Exploration Satellite (Earth-to-space), secondary, and Meteorological Satellite (Earth-to space), secondary, shared.
406-406.1 MHz:	Mobile Satellite (Earth-to-space), exclusive.
406.1-410 MHz:	Radio Astronomy, primary, shared.
460-470 MHz:	Meteorological Satellite (space-to-Earth), secondary, shared.
608-614 MHz:	Region 2, Radio Astronomy, primary; Mobile Satellite except aeronautical mobile satellite (Earth-to-space), secondary, shared.
1 215-1 240 MHz:	Radionavigation Satellite (space-to-Earth), primary, shared.
1 240-1 260 MHz:	Radionavigation Satellite (space-to-Earth), primary, shared.
1 400-1 427 MHz:	Earth Exploration Satellite (passive), primary, Radio Astronomy, primary, and Space Research (passive), primary, shared.
1 427-1 429 MHz:	Space Operation (Earth-to-space), primary, shared.
1 525-1 530 MHz:	Space Operation (space-to-Earth), primary, and Earth Exploration Satellite, secondary, shared differently in regions.

1 530-1 535 MHz:	Space Operation (space-to-Earth), primary, Maritime Mobile Satellite (space-to-Earth), primary, and Earth Exploration Satellite, secondary, shared differently in each region.
1 535-1 544 MHz:	Maritime Mobile Satellite (space-to-Earth), exclusive.
1 544-1 545 MHz:	Mobile Satellite (space-to-Earth), exclusive.
1 545-1 559 MHz:	Aeronautical Mobile Satellite (R) (space-to-Earth), exclusive.
1 559-1 610 MHz:	Radionavigation Satellite (space-to-Earth), primary, shared.
1 626.5-1 645.5 MHz:	Maritime Mobile Satellite (Earth-to-space), exclusive.
1 645.5-1 646.5 MHz:	Mobile Satellite (Earth-to-space), exclusive.
1 646.5-1 660 MHz:	Aeronautical Mobile Satellite (R) (Earth-to-space), exclusive.
1 660-1 660.5 MHz:	Aeronautical Mobile Satellite (R) (Earth-to-space), primary, and Radio Astronomy, primary.
1 660.5-1 668.4 MHz:	Radio Astronomy, primary, and Space Research (passive), primary, shared.
1 668.4-1 670 MHz:	Radio Astronomy, primary, shared.
1 670-1 690 MHz:	Meteorological Satellite (space-to-Earth), primary, shared.
1 690-1 700 MHz:	Meteorological Satellite (space-to-Earth), primary, shared differently in regions 1, and 2 and 3.
1 700-1 710 MHz:	Meteorological Satellite (space-to-Earth), primary, shared differently in regions 1, and 2 and 3.

2 290-2 300 MHz:	Space Research (deep space) (space to Earth), primary, shared differently in regions 1, and 2 and 3.
2 500-2 655 MHz:	a. Region 1, Broadcasting Satellite, primary, shared.
	b. Region 2, Fixed Satellite (space-to-Earth), primary and Broadcasting Satellite, primary, shared.
	c. Region 3, 2 500-2 535 MHz., Fixed Satellite (space-to-Earth), primary, and Broadcasting Satellite, primary, shared.
	d. Region 3, 2 535-2 655 Mhz., Broadcasting Satellite, primary, shared.
2 655-2 690 MHz:	a. Region 1, Broadcasting Satellite, primary, Earth Exploration Satellite (passive), secondary, Radio Astronomy, secondary, and Space Research (passive), secondary, shared.
	b. Region 2, Fixed Satellite (Earth-to-space) (space-to-Earth), primary, Broadcasting Satellite, primary, Earth Exploration Satellite (passive), secondary, Radio Astronomy, secondary, Space Research (passive), secondary, shared.
	c. Region 3, Fixed Satellite (Earth-to-space), primary, Broadcasting Satellite, primary, Earth Exploration Satellite (passive), secondary, Radio Astronomy, secondary, and Space Research (passive), secondary, shared.
2 690-2 700 MHz:	Earth Exploration Satellite (passive), primary, Radio Astronomy, primary, and Space Research (passive), primary.
3 400-4 200 MHz:	a. Region 1, 3 400-3 600 MHz and 3 600-4 200 MHz, Fixed Satellite (space-to-Earth), primary, shared differently.

b. Regions 2 & 3, 3 400-3 500 MHz, 3 500-3 700 MHz, and 3 700-4 200 MHz, Fixed Satellite (space-to-Earth), primary, shared differently.

4 500-4 800 MHz: Fixed Satellite (space-to-Earth), primary, shared.

4 800-4 990 MHz: Radio Astronomy, secondary, shared.

4 990-5 000 MHz: Radio Astronomy, primary, and Space Research (passive), secondary, shared.

5 250-5 255 MHz: Space Research, secondary, shared.

5 650-5 725 MHz: Space Research (deep space), secondary, shared.

5 725-5 850 MHz: Region 1, Fixed Satellite (Earth-to-space), primary, shared.

5 850-5 925 MHz: Fixed Satellite (Earth-to-space), primary, shared differently in regions.

5 925-7 075 MHz: Fixed Satellite (Earth-to-space), primary, shared.

7 250-7 300 MHz: Fixed Satellite (space-to-Earth), primary, shared.

7 300-7 450 MHz: Fixed Satellite (space-to-Earth), primary, shared.

7 450-7 550 MHz: Fixed Satellite (space-to-Earth), primary, and Meteorological Satellite (space-to-Earth), primary, shared.

7 550-7 750 MHz: Fixed Satellite (space-to-Earth), primary, shared.

7 900-7 975 MHz: Fixed Satellite (Earth-to-space), primary, shared.

7 975-8 025 MHz: Fixed Satellite (Earth-to-space), primary, shared.

8 025-8 175 MHz: a. Regions 1 and 3, Fixed Satellite (Earth-to-space), primary, and Earth Exploration Satellite (space-to-Earth), secondary, shared.

b. Region 2, Earth Exploration Satellite (space-to-Earth), primary, and Fixed Satellite (Earth-to-space), primary, shared.

8 175-8 215 MHz: a. Region 1, Fixed Satellite (Earth-to-space), primary, Meteorological Satellite (Earth-to-space), primary, and Earth Exploration Satellite (space-to-Earth), secondary, shared.

b. Region 2, Earth Exploration Satellite (space-to-Earth), primary, Fixed Satellite (Earth-to-space), primary, and Meteorological Satellite (Earth-to-space), primary, shared.

c. Region 3, Fixed Satellite (Earth-to-space), primary, Meteorological Satellite (Earth-to-space), primary, and Earth Exploration Satellite (space to Earth), secondary, shared.

8 215-8 400 MHz: a. Regions 1 and 3, Fixed Satellite (Earth-to-space), primary, and Earth Exploration Satellite (space-to-Earth), secondary, shared.

b. Region 2, Earth Exploration Satellite (space-to-Earth), primary, and Fixed Satellite (Earth-to-space), primary, shared.

8 400-8 500 MHz: Space Research (space to Earth), primary, shared.

10.45-10.5 GHz: Amateur Satellite, secondary, shared.

10.6-10.68 GHz:	Earth Exploration Satellite (passive), primary, Radio Astronomy, primary, and Space Research (passive), primary,
10.68-10.7 GHz:	Earth Exploration Satellite (passive), primary,
	Radio Astronomy, primary, and Space Research (passive), primary.
10.7-11.7 GHz:	a. Region 1, Fixed Satellite (space-to-Earth) (Earth-to-space), primary, shared.
	b. Regions 2 and 3, Fixed Satellite (space-to-Earth), primary, shared.
11.7-12.75 GHz:	a. Region 1, 11.7-12.5 GHz, Broadcasting Satellite, primary, shared, and 12.5-12.75 GHz., Fixed Satellite (space-to-Earth and Earth-to-space), exclusive.
	b. Region 2, 11.7-12.1 GHz, Fixed Satellite (space-to-Earth), primary, shared; 12.1-12.2 GHz, Fixed Satellite (space-to-Earth), exclusive, 12.2-12.7 GHz, Broadcasting Satellite, primary, shared and 12.7-12.75 GHz, Fixed Satellite (Earth-to-space), primary, shared.
	c. Region 3, 11.7-12.2 GHz, Broadcasting Satellite, primary, shared, and 12.5-12.75 GHz, Fixed Satellite (space-to-Earth), primary, and Broadcasting Satellite, primary, shared.
12.75-13.25 GHz:	Fixed Satellite (Earth-to-space), primary, and Space Research (deep space)(space-to-Earth), secondary, shared.
13.4-14 GHz:	Standard Frequency and Time Signal Satellite (Earth-to-space), secondary, and Space Research, secondary, shared.
14-14.25 GHz:	Fixed Satellite (Earth-to-space), primary, and Space Research, secondary, shared.

14.25-14.3 GHz: Fixed Satellite (Earth-to-space), primary, and Space Research, secondary, shared.

14.3-14.4 GHz: Fixed Satellite (Earth-to-space), primary, and Radionavigation Satellite, secondary, shared in Regions 1 and 3, and not shared in Region 2.

14.4-14.47 GHz: Fixed Satellite (Earth-to-space), primary, and Space Research (space-to-Earth), secondary, shared.

14.47-14.5 GHz: Fixed Satellite (Earth-to-space), primary, and Radio Astronomy, secondary, shared.

14.5-14.8 GHz: Fixed Satellite (Earth-to-space), primary, and Space Research, secondary, shared.

14.8 15.35 GHz Space Research, secondary, shared.

15.35-15.4 GHz: Earth Exploration Satellite (passive), primary, and Space Research (passive), primary, shared.

16.6-17.1 GHz: Space Research (deep space)(Earth-to-space), secondary, shared.

17.2-17.3 GHz: Earth Exploration Satellite (active), secondary, and Space Research (active), secondary, shared.

17.3-17.7 GHz: Fixed Satellite (Earth-to-space), primary, shared.

17.7-18.1 GHz: Fixed Satellite (space-to-Earth and (Earth-to-space), primary, shared.

18.1-18.6 GHz: Fixed Satellite (space-to-Earth), primary, shared.

18.6-18.8 GHz: a. Regions 1 and 3, Fixed Satellite (space-to-Earth), primary, Earth Exploration Satellite (passive), secondary, and Space Research (passive), secondary, shared.

b. Region 2, Earth Exploration Satellite (passive) primary, Fixed Satellite (space-to-Earth), primary, and Space Research (passive), primary, shared.

18.8-19.7 GHz: Fixed Satellite (space-to-Earth), primary, shared.

19.7-20.2 GHz: Fixed Satellite (space-to-Earth), primary, and Mobile Satellite (space-to-Earth), secondary.

20.2-21.2 GHz: Fixed Satellite (space-to-Earth), primary, Mobile Satellite (space-to-Earth), primary, and Standard Frequency and Time Signal Satellite (space-to-Earth), secondary.

21.2-21.4 GHz: Earth Exploration Satellite (passive), primary, Space Research (passive), primary, shared.

22.21-22.5 GHz: Earth Exploration Satellite (passive), primary, Radio Astronomy, primary, and Space Research (passive), primary, shared.

22.5-22.55 GHz: Regions 2 and 3, Broadcasting Satellite, primary, shared.

22.55-23 GHz: a. Region 1, Inter-Satellite, primary, shared.

b. Regions 2 and 3, Inter-Satellite, primary, and Broadcasting Satellite, primary, shared.

23-23.55 GHz: Inter-Satellite, primary, shared.

23.6-24 GHz: Earth Exploration Satellite (passive), primary, Radio Astronomy, primary, Space Research (passive), primary.

24-24.05 GHz: Amateur Satellite, primary, shared.

24.05-24.25 GHz: Earth Exploration Satellite (active), secondary, shared.

25.25-27 GHz: Earth Exploration Satellite (space-to-space), secondary, and Standard Frequency and Time

Signal Satellite (Earth-to-space), secondary, shared.

27-27.5 GHz: a. Region 1, Earth Exploration Satellite (space-to-space), secondary, shared.

b. Regions 2 and 3, Fixed Satellite (Earth-to-space), primary and Earth Exploration Satellite (space-to-space), secondary, shared.

27.5-29.5 GHz: Fixed Satellite (Earth-to-space), primary, shared.

29.5-30 GHz: Fixed Satellite (Earth-to-space), primary, and Mobile Satellite (Earth-to-space), secondary.

30-31 GHz: Fixed Satellite (Earth-to-space), primary, Mobile Satellite (Earth-to-space), primary, and Standard Frequency and Time Signal Satellite (space-to-Earth), secondary.

31-31.3 GHz: Standard Frequency and Time Signal Satellite (space-to-Earth), secondary, and Space Research, secondary, shared.

31.3-31.5 GHz: Earth Exploration Satellite (passive), primary, Radio Astronomy, primary, and Space Research (passive), primary.

31.5-31.8 GHz: Earth Exploration Satellite (passive), primary, Radio Astronomy, primary, and Space Research (passive), primary, shared in Regions 1 and 3, and not shared in Region 2.

31.8-32 GHz: Space Research, secondary, shared.

32-32.3 GHz: Inter-Satellite, primary, and Space Research, secondary, shared.

32.3-33 GHz: Inter-Satellite, primary, shared.

34.2-35.2 GHz: Space Research, secondary, shared.

36-37 GHz:	Earth Exploration Satellite (passive), primary, and Space Research (passive), primary, shared.
37.5-39.5 GHz:	Fixed Satellite (space-to-Earth), primary, shared.
39.5-40.5 GHz:	Fixed Satellite (space-to-Earth), primary, and Mobile Satellite (space-to-Earth), primary, shared.
40.5-42.5 GHz:	Broadcasting Satellite, primary, shared.
42.5-43.5 GHz:	Fixed Satellite (Earth-to-space), primary, and Radio Astronomy, primary, shared.
43.5-47 GHz:	Mobile Satellite, primary, and Radionavigation Satellite, primary, shared.
47-47.2 GHz:	Amateur Satellite, primary, shared.
47.2-50.2 GHz:	Fixed Satellite (Earth-to-space), primary, shared.
50.2-50.4 GHz:	Earth Exploration Satellite (passive), primary, and Space Research (passive), shared.
50.4-51.4 GHz:	Fixed Satellite (Earth-to-space), primary, and Mobile Satellite (Earth-to-space), secondary, shared.
51.4-54.25 GHz:	Earth Exploration Satellite (passive), primary, and Space Research (passive), primary.
54.25-58.2 GHz:	Earth Exploration Satellite (passive), primary, Inter-Satellite, primary, and Space Research (passive), primary, shared.
58.2-59 GHz:	Earth Exploration Satellite (passive), primary, and Space Research (passive) primary.
59-64 GHz:	Inter-Satellite, primary, shared.
64-65 GHz:	Earth Exploration Satellite (passive), primary, and Space Research (passive), primary.

65-66 GHz:	Earth Exploration Satellite, primary, and Space Research, primary, shared.
66-71 GHz:	Mobile Satellite, primary, and Radionavigation Satellite, primary, shared.
71-74 GHz:	Fixed Satellite (Earth-to-space), primary, and Mobile Satellite (Earth-to-space), primary, shared.
74-75.5 GHz:	Fixed Satellite (Earth to space), primary, shared.
75.5-76 GHz:	Amateur Satellite, primary, shared.
76-81 GHz:	Amateur Satellite, secondary, shared.
81-84 GHz:	Fixed Satellite (space-to-Earth), primary, and Mobile Satellite (space-to-Earth), primary, shared.
84-86 GHz:	Broadcasting Satellite, primary, shared.
86-92 GHz:	Earth Exploration Satellite (passive), primary, Radio Astronomy, primary, and Space Research (passive), primary.
92-95 GHz:	Fixed Satellite (Earth-to-space), primary, shared.
95-100 GHz:	Mobile Satellite, primary, and Radionavigation Satellite, primary, shared.
100-102 GHz:	Earth Exploration Satellite (passive), primary, and Space Research (passive), primary, shared.
102-105 GHz:	Fixed Satellite (space-to-Earth), primary, shared.
105-116 GHz:	Earth Exploration Satellite (passive), primary, Radio Astronomy, primary, and Space Research (passive), primary.

116-126 GHz:	Earth Exploration Satellite (passive), primary, Inter-Satellite, primary, and Space Research (passive), primary, shared.
126-134 GHz:	Inter-Satellite, primary, shared.
134-142 GHz:	Mobile Satellite, primary, and Radionavigation Satellite, primary, shared.
142-144 GHz:	Amateur Satellite, primary, shared.
144-149 GHz:	Amateur Satellite, secondary, shared.
149-150 GHz:	Fixed Satellite (space-to-Earth), primary, shared.
150-151 GHz:	Earth Exploration Satellite (passive), primary, Fixed Satellite (space-to-Earth), primary, and Space Research (passive), primary, shared.
151-164 GHz:	Fixed Satellite (space-to-Earth), primary, shared.
164-168 GHz:	Earth Exploration Satellite (passive), primary, Radio Astronomy, primary, and Space Research (passive), primary.
170-174.5 GHz:	Inter-Satellite, primary, shared.
174.5-176.5 GHz:	Earth Exploration Satellite (passive), primary, Inter-Satellite, primary, and Space Research (passive), primary, shared.
176.5-182 GHz:	Inter-Satellite, primary, shared.
182-185 GHz:	Earth Exploration Satellite (passive), primary, Radio Astronomy, primary, and Space Research (passive), primary.
185-190 GHz:	Inter-Satellite, primary, shared.
190-200 GHz:	Mobile Satellite, primary, and Radionavigation Satellite, primary, shared.

200-202 GHz:	Earth Exploration Satellite (passive), primary, and Space Research (passive), primary, shared.
202-217 GHz:	Fixed Satellite (Earth-to-space), primary, shared.
217-231 GHz:	Earth Exploration Satellite (passive), primary, Radio Astronomy, primary, and Space Research (passive), primary.
231-235 GHz:	Fixed Satellite (space-to-Earth), primary, shared.
235-238 GHz:	Earth Exploration Satellite (passive), primary, Fixed Satellite (space-to-Earth), primary, and Space Research (passive), primary, shared.
238-241 GHz:	Fixed Satellite (space-to-Earth), primary, shared.
241-248 GHz:	Amateur Satellite, secondary, shared.
248-250 GHz:	Amateur Satellite, primary, shared.
250-252 GHz:	Earth Exploration Satellite (passive), primary, and Space Research (passive), primary.
252-265 GHz:	Mobile Satellite, primary, and Radionavigation Satellite, primary, shared.
265-275 GHz:	Fixed Satellite (Earth-to-space), primary, and Radio Astronomy, primary, shared.

APPENDIX 2

1989 SATELLITE PERFORMANCE CHART*

* Published with the permission of *Satellite Communications*, 6300 S. Syracuse Way, Suite 650, Englewood, Colorado, 80111.

NORTH AMERICAN DOMESTIC SATELLITES

Name	Operator	Launch Dates	Transponders per Satellite	Bandwidth (MHz)	Uplink (GHz)	G/T (dBi/K) @ edge	Downlink (GHz)	eirp (dBW) @ edge	Lifetime (years)	Orbit Location (longitude)
Anik C1 to C2	Telesat Canada	85,83,82	16	864	14.0-14.5	+3	11.7-12.2	47	10	107.5,110,117.5W
Anik D1 to D2	Telesat Canada	82,84	24	864	5.925-6.425	0	3.7-4.2	38	8-10	104.5,111.5W
Anik E1 to E2	Telesat Canada	90	24	864	5.925-6.425	0	3.7-4.2	40	10	107.5,110.5W
			16	864	14.0-14.5	+3	11.7-12.2	47		
Aurora I	Alascom, Inc.	82	24	864	5.925-6.425	+4	3.7-4.2	32-38	10	143W
Aurora II	Alascom, Inc.	91	24	864	5.925-6.425	+4	3.7-4.2	34-40	12	137W
ASC 1 & 2	Contel-ASC	85,90	18	864	5.925-6.425	-1	3.7-4.2	34 & 36	8-10	129,83W
			6	432	14.0-14.5	-4	11.7-12.2	40		
Contelsat -1,-2,-3	Contel-ASC	93,93,Spare	16	864	14.0-14.5	-4	11.7-12.2	35	10	101,129W
			24	864	5.925-6.425	+2	3.7-4.2	46-49		
Galaxy I-IV	Hughes Comm., Galaxy, Inc.	83,83,84,91	24	864	5.925-6.425	-2	3.7-4.2	34.5	9-12	133,74,95,141W
Galaxy V-VI	Hughes Comm., Galaxy, Inc.	93,91	24	864	5.925-6.425	-2	3.7-4.2	34.5	9-12	64.91W
Galaxy IR,IIR,IIIR	Hughes Comm., Galaxy, Inc.	93,94,94	24	864	5.925-6.425	-2	3.7-4.2	TBD	10	133,74,95W
Galaxy A & B	Hughes Comm., Galaxy, Inc.	93,94	24	864	14.0-14.5	+2	11.7-12.2	47-49.5	12	99,131W
Gstar I-IV	GTE Spacenet Corp.	85,86,88,90	16	864	14.0-14.5	-2	11.7-12.2	38-48	10(a)	121,105,125,64W
Gstar IR	GTE Spacenet Corp.	94	24	864	14.0-14.5	1	11.7-12.2	43.8	10	121W
Morelos A & B	Mexico	85,85	18	864	5.925-6.425	+2	3.7-4.2	36-39	10	113.5,116.5W
			4	432	14.0-14.5	+0.1	11.7-12.2	44		

Satellite	Operator	Orbital Position								Longitude
Satcom IR,IIR	GE Americom	83,83	24	864	5.925-6.425	-6	3.7-4.2	34	10	139.72W
Satcom IIIR & IV	GE Americom	81,82	24	864	5.925-6.425	-4	3.7-4.2	34 & 32	11	131.81W
Satcom C-1(b)	GE Americom	93	24	864	5.925-6.425	-4	3.7-4.2	34	TBD	138W
Satcom C-3,C-4(c)	GE Americom	92,93	24	864	5.925-6.425	-6	3.7-4.2	34 & 32	12	130.81W
Satcom K1-K2	GE Americom	86,85	16	864	14.0-14.5	0	11.7-12.2	43-49	10	67.81W
Satcom K3	Crimson Associates	89	16	864	14.0-14.5	+2	11.7-12.2	45-49	10	85W
Satcom H-1	GE Americom	93	24	864	5.925-6.425	-2	3.7-4.2	34	10	79W
SBS 1-2	Comsat General	80,81	16	864	14.0-14.5	+2	11.7-12.2	42	7(d)	74W
SBS 3	MCI	82	10	430	14.0-14.5	-2	11.7-12.2	41	8-9(d)	95W
SBS 4	IBM/Hughes	84	10	430	14.0-14.5	-2	11.7-12.2	41-43	9	91W
SBS 5	IBM/Hughes	88	14	430	14.0-14.5	-5	11.7-12.2	40	10	123W
SBS 6	IBM/Hughes	90	19	870	14.0-14.5	0	11.7-12.2	40	10	72W
Spacenet I-IV	GTE Spacenet Corp.	84,84,88,90	18	817	5.925-6.425	-4	3.7-4.2	42	10	120,69,87,141W
Spacenet IR,IIR, IIR	GTE Spacenet Corp.	93,93, Spare	6	864	5.925-6.425	-2	3.7-4.2	34 & 36	10	103.69W
Spotnet 1, 2, 3	National Exchange Satellite, Inc.	93,93, Spare	24	432	5.925-6.425	-1	3.7-4.2	39	10	135.76W
Telstar 301-303	AT&T Communications	83,84,85	18	864	14.0-14.5	-5.9	11.7-12.2	35	10	97,85,123W
Telstar 401,402, 403	AT&T Communications	92,93, Spare	24	3753	14.0-14.5	+5	3.7-4.2	40.8	10	97,89W
Westar III	Hughes Comm. Galaxy, Inc.	79	24	864	5.925-6.425	-5	3.7-4.2	31	8	91W
Westar IV & V	Hughes Comm. Galaxy, Inc.	82,82	12	432	5.925-6.425	-7.4	11.7-12.2	56-59	10	99,122.5W

(a)Except Gstar III, which is 3 years.   (c)Will replace Satcom IIIR and IV in 1992 and 1993, respectively.

(b)Will replace Satcom IR in 1993.   (d)Lifetime may be extended by using the COMSAT Maneuver.

EXISTING AND NEAR-FUTURE FOREIGN SATELLITES

Name	Operator	Launch Dates	Transponders per Satellite	Bandwidth (MHz)	Uplink (GHz)	G/T (dB/K) @ edge	Downlink (GHz)	eirp (dBW) @ edge	Lifetime (years)	Orbit Location (longitude)
ASIA										
Asiasat 1 (Alias Westar VI)	Asia Satellite Telecommunications	94	24	864	5.925-6.425	-1	3.7-4.2	34	12	TBD
BS 2b (Yuri 2b)	MOPT, Japan	86	2	140	14.0-14.5	+2.4	11.7-12.2	46-55	3	110E
China Sat	P.R. China (PRC)	86	1	36	6	TBD	4	50	3	103E
CS-2a & -2b (Sakura 2)	NASDA-NTT (Japan)	83,83	2	360	5.925-6.425	-8	3.7-4.2	29	5	127 & 136E
			6	780	27.55-30.05	-5	17.75-20.25	37		
CS-3a & 3b (Sakura 3)	NASDA-NTT (Japan)	92,94	2	360	5.925-6.425	TBD	3.7-4.2	TBD	10	TBD
			10	1300	27.55-30.05	TBD	17.75-20.25			
Insat IB & IC	Indian Space Res Org	83,88	2	72	5.855-5.935	-6	2.555-2.635	42	7	74,94E
			12	432	5.935-6.425	-6	3.710-4.200	32		
Insat II	Indian Space Res Org	93+	24	432	5.850-6.425	-6	3.7-4.2	32	7	TBD
			2	432	5.9	-6	2.5	42		
JC Sat-1, -2	Japan Comm Satellite	89,90	32	864	14.0-14.5	+6	12.25-12.75	48	10-12	150,154E
Palapa B1, B2-P, B2-R	Perumtel (Indonesia)	83,87,90	24	864	5.925-6.425	-5	3.7-4.2	34	8	108,113,118E
Superbird	Space Comm Corporation (Japan)	89	12	432	14.0-14.5	TBD	12.5-12.75	TBD	10	124,128E
			23	TBD	27.55-30.05	TBD	17.75-20.25	TBD		
AUSTRALIA										
Aussat 1 to 3	Aussat Pty Ltd	85,85,87	15	675	14.0-14.5	-3	12.25-12.75	36-47	8	156,160,164E
Aussat B1 to B3	Aussat Pty Ltd	92,93,97	15	810	14.0-14.5	TBD	12.25-12.75	45	10	156,160,164E
			1	14	1.6	TBD	1.5	TBD		

EUROPE

Satellite	Operator									
Astra 1A	Soc. Europeen) des Satellites (SES)	88	16	400	14.25-14.5	+3	11.2-11.45	50	10	19.2E
Astra 1B	Soc. Europeene des Satellites (SES)	93	16	400	14.25-14.5	+3	11.2-11.45	50	10	TBD
Atlantic Sat	Hughes/Ireland	90	24	1296	12.75-13.25	+9	10.7-11.2	48	10	31W
			5	135	17	+9	11.746-12.054	61.5		
BSB	British Sat. Broad.	89+	5	135	17	+7	11.785-12.015	62	10	31W
DFS (Kopernikus)	Deutsche Bundespost	90	3	270	14.0-14.25	9.6	12.5-12.75	49	10	23.5E
			7	308	30	7.7	20	48		
			1	90		8.9	11.450-11.700	49		
Ekran-M (Statsionar-T)	USSR	87+	1	25	6.2	-19	0.714-0.754	50	1-2	99E
Eutelsat I (ESC-I)	Eutelsat	83,84,87	12-14	800	14.0-14.5	-5.3	10.95-11.70	35-41	10	13. 7 & 16E
						-3	12.5-12.75	41		
ECS 1, 2 & 4		90+	14	1250	14.0-14.5	-5.3	10.95-11.70	45-51	10	36. 3 & 19E
Eutelsat II (ESC-II)	Eutelsat					-3	12.5-12.75	45		
Italsat 1, 2	Italy	90,92	9	828	30	6-16	20	42-45	10	13E
			2	54	17.7	+6	12.5	63	5	19W
Olympus	ESA	89	3	780	30	TBD	20	52		
			4	144	13 & 14	1.1	12.507-12.598	44		
TDF-1, 2	France	88+	5	135	17.3	+5	12.1	60	9	19W
TV-SAT-1, 2	Germany	87+	5	135	17.7	+7	12.0	60	7	19W
Sarit 1, 2	Italy	91,92	5	135	17.7	TBD	12.0	57	10	19W
Telecom I	Telecom France	83+	2	80	5.925-6.420	-16	3.7-4.195	28-35	7	8.5.3W
			2	240	5.925-6.420	-15	3.7-4.195	26		
			2	72	7.98-8.095	-13	7.255-7.37	33		
			6	216	14.0-14.25	+8	12.5-12.75	44		
Telecom II	Telecom France	91,92,94	10	TBD	5.925-6.425	-16	3.7-4.2	TBD	10	3.5.8W
			5	360	8	-13	7	26		
Tele-X	Notelsat	89	11	396	14.0-14.25	+8	12.5-12.75	46	5	5E
			2	126	14.0-14.5	TBD	12.5-12.75	60		
			2	54	17.7-18.1	TBD	11.7-12.5	TBD		

(Cont.)

EXISTING AND NEAR-FUTURE FOREIGN SATELLITES (Cont.)

Name	Operator	Launch Dates	Transponders per Satellite	Bandwidth (MHz)	Uplink (GHz)	G/T (dBi/K) edge ⓓ	Downlink (GHz)	eirp (dBW) edge ⓓ	Lifetime (years)	Orbit Location (longitude)
MIDDLE EAST										
AMS-1,-2 (AMOS)	Israel	92,93	17	612	5.85-6.55	+7	3.5-4.2	41	TBD	15E
Arabsat IA, IB	Arab. Sat. Comm Consortium	85+	12	432	12.75-13.25	+11	11.2-11.7	38-45	7	19, 26E
			25	825	5.925-6.435	-6	3.7-4.2	31		
			1	33	5.925-6.435	-6	2.54-2.65	41		
SOUTH AMERICA										
Brasilsat	Embratel*	85,86	24	864	5.925-6.425	+2 to -4	3.7-4.2	33	8	65, 70W

*Empresa Brasileira de Telecomunicacoes.

INTERNATIONAL SATELLITES IN THE FIXED SATELLITE SERVICE (FSS)

Name	Operator	Launch Dates	Transponders per Satellite	Bandwidth (MHz)	Uplink (GHz)	G/T (dBi/K) edge	Downlink (GHz)	eirp (dBW) edge	Lifetime (years)	Orbit Location (longitude)
Intelsat V (F-1 to F-8)	Intelsat	80+	4	136	5.925-6.425	-14.6	3.7-4.2	23-9	7	63,66,174,180E
			5	180	5.925-6.425	-14.6&-21.6	3.7-4.2	20-26		1,18.5,21.5,34.5
						-21.6				50.53W
			1	41	5.925-6.425	-21.6	3.7-4.2	20.5		
			10	720	5.925-6.425	-11.6&-14.6	3.7-4.2	23-26		
			4	308	5.925-6.425	-11.6&-14.6	3.7-4.2	23-26		
			2	144	14.169-14.241	-3 & 0	10.95-11.7	38-41		
			2	154	14.004-14.081	-3 & 0	10.95-11.7	38-41		
			2	482	14.259-14.498	-3 & 0	10.95-11.7	38-41		
Intelsat VA (F-10 to F-12)	Intelsat	85+	10	360	5.925-6.425	-14.6	3.7-4.2	20-26	7	60,174,177E
			2	82	5.925-6.425	-21.6	3.7-4.2	20-26		1,24.5,27.5,34.5W
			13	936	5.925-6.425	-11.6&-14.6	3.7-4.2	20-26		
			4	308	5.925-6.425	-11.6&-14.6	3.7-4.2	23-26		
			2	144	14.169-14.241	-3 & 0	10.95-11.7	38-41		
			2	154	14.004-14.081	-3 & 0	10.95-11.7	38-41		
			2	482	14.259-14.498	-3 & 0	10.95-11.7	38-41		
Intelsat VA (IBS)	Intelsat	88+	10	360	5.925-6.425	-11.6&-14.6	3.7-4.2	20-26	7	18.5, 53W
			2	82	5.925-6.425	-11.6&-14.6	3.7-4.2	20-26		
			13	936	5.925-6.425	-11.6&-14.6	3.7-4.2	20-26		
			4	308	5.925-6.425		3.7-4.2	23-26		
			4	288	14.0-14.5	-3 & -8	10.95-11.7	38-41		
			2	154	14.0-14.5	-3 & -8	10.95-11.7	38-41		
			2	482	14.0-14.5	-3 & -8	10.95-11.7	38-41		
			4	288	14.0-14.5	-3	11.7-11.95	TBD		
			4	154	14.0-14.5	-3	11.7-11.95	TBD		
			4	288	14.0-14.5	-8	12.5-12.75	TBD		
			2	154	14.0-14.5	-8	12.5-12.75	TBD		

(Cont.)

INTERNATIONAL SATELLITES IN THE FIXED SATELLITE SERVICE (FSS) (Cont.)

Name	Operator	Launch Dates	Transponders per Satellite	Bandwidth (MHz)	Uplink (GHz)	G/T (dB/K) @ edge	Downlink (GHz)	eirp (dBW) @ edge	Lifetime (years)	Orbit Location (longitude)
Intelsat VI (F-1 to F-5)	Intelsat	89+	12	432	5.85-6.425	-21.6&-11.6	3.625-4.2	20-26	10	60, 63E
			2	82	5.85-6.425	-21.6&-11.6	3.625-4.2	20-26		24.5,27.5,34.5E
			28	1872	5.85-6.425	-21.6&-11.6	3.625-4.2	23-26		
			4	288	14.0-14.5	TBD	10.95-11.7	TBD		
			2	154	14.0-14.5	TBD	10.95-11.7	TBD		
			2	300	14.0-14.5	TBD	10.95-11.7	TBD		
Intelsat VII (F-1 to F-5)	Intelsat	92-94	28	1512	5.85-6.425	-21.6&-11.6	3.625-4.2	26-36	TBD	174,177,307,325.5,
			10	880	14.0-14.5	-21.6&-11.6	10.95-12.75	41-46		342,359E

OTHER INTERNATIONAL SATELLITES

Name	Operator	Launch Dates	Transponders per Satellite	Bandwidth (MHz)	Uplink (GHz)	G/T (dB/K) @ edge	Downlink (GHz)	eirp (dBW) @ edge	Lifetime (years)	Orbit Location (longitude)
Celestar I & II	McCaw Communications	94.95	102	3672	14.0-14.5	+12	11.7-12.2	46-48	11	170,70E
Finansat 1 & 2	Financial Satellite International	93.94	24	864	5.925-6.425	0	3.7-4.2	37.6	10	47, 178W
ISI	Satellite, Inc.	TBD	34	1836	14.0-14.5	+2 to +5	10.7-10.9	42-44	10	56, 58W
Orion	Orion Satellite Corporation	TBD	44	432	14.0-14.25	TBD	11.45-11.7	TBD	TBD	37.5,40,47W
				864	14.0-14.50	TBD	11.7-12.2	TBD		
				432	14.25-14.5	TBD	12.5-12.75			
Pacstar-A & -B	Pacific Satellite Inc.	90	TBD	TBD		+5 to +8	1.545-1.559	TBD	10	167E, 175W
			17	TBD	1.640-1.660	+5	3.61-4.19	34-38		
			4	TBD	6.495-7.075	+5	11.7-12.2	43.5		
			4	TBD	14.0-14.5			43.5		
PAS-1	PanAmSat	88	18	864	5.925-6.425	-11	12.5-12.75	31	8.5	45W
			6	432	14.0-14.5	-4	11.48-11.92	43		

INTERNATIONAL SATELLITES IN THE FIXED SATELLITE SERVICE IN GOVERNMENT USE

Name	Operator	Launch Dates	Transponders per Satellite	Bandwidth (MHz)	Uplink (GHz)	G/T (dB/K) @ edge	Downlink (GHz)	eirp (dBW) @ edge	Lifetime (years)	Orbit Location (longitude)
DSCS-III	US Air Force	82+	1	395	7.9-8.4	-15	7.25-7.75	25	10	60 & 175E
			4	TOTAL	7.9-8.4	-16.-1	7.25-7.75	23-44	10	13 & 135W
			1		7.9-8.4	-15	7.25-7.75	23-37.5		
					0.3-0.4	-17	0.225-0.26			
Fltsatcom	US Navy	78+	12	0.7	0.29-0.32	-16.7	0.24-0.27	26	5	23,93 & 100W
			1	TOTAL	0.29-0.32	-16.7	0.24-0.27	16.5		72.5, 172E
			1		0.29-0.32	-16	0.24-0.27	26-28		
								27		
Leasat	Hughes Comm Serv. Inc. (for US Navy)	84+	9	0.7	0.29-0.32	-16	0.24-0.27	26, 28	10	105, 15, 23W
			5	TOTAL	0.29-0.32	-8	0.24-0.27	6.5		
			6		0.29-0.32	-18	0.24-0.27	26		
			1*		8	-18	0.24-0.27	28		
						-20		26		
Milstar	US Dept of Defense	89+	TBD	TBD	44	TBD	20	TBD	10	TBD
NATO III	NATO	76+	1	152	7.95-8.162	-14	7.25-7.437	31 & 37	7	30,18,13.5,52W
			1	TOTAL	7.95-8.162	-14	7.25-7.437	31 & 36		
			1		7.95-8.162	-14	7.25-7.437	31 & 36		
TDRS	Contel (for NASA)	83+	1 (MA)	331	2.2875	-13.7	2.10	26.3	TBD	41, 171W
			2 (SSA)	TOTAL	2.2-2.3	+8.9	2.0-2.1	44.0-47.7		
			2 (KSA)		15	+24.6	13.8	47.4-52.2		
UHF Follow-On	US Navy	92+	TBD	TBD	0.29-0.32	TBD	0.24-0.27	TBD	14	TBD

*Fleet broadcast.

COMMERCIAL MOBILE SATELLITES

Name	Operator	Launch Dates	Transponders per Satellite	Bandwidth (MHz)	Uplink (GHz)	G/T (dB/K) @ edge	Downlink (GHz)	eirp (dBW) @ edge	Lifetime (years)	Orbit Location (longitude)
Geostar (on GTE)	Geostar Corporation	87	1	TBD	1.61-1.627	TBD	12	TBD	See GTE Gstar & Spacenet	125, 87W
Geostar RDSS	Geostar Corporation	91+	4	16.5	1.61-1.627	TBD	2.483-2.500	TBD	10	100,70,130W
				16.5	6.525-6.542	TBD	5.142-5.225			
INMARSAT 2	Inmarsat	91+	1	TBD	1.6	-15.0	1.5	TBD	10	15.26W,64.5E 60,63,66,179E
Maritime Comm. Subsystem (MCS)	Intelsat & Inmarsat	82+	1	7.5	1.6	-15.0	1.5	27	7	18.5, 21.5E (on Intelsat Va)
Marecs A & B	ESA & Inmarsat	81+	1 ea.	5.5	1.6	-13.3	1.5	34	7	27.9W,177.8,64.5E
Marisat	Comsat General & Inmarsat	76+	1 ea.	4	1.6	-17	1.5	30-33	TBD	73,176.5E,15W
AMSC & M-SAT	(US & Canada)	93+	TBD	TBD	1.6	+3	1.5	56	7-10	106.5,101,62,137W
			TBD	200	13.0-13.2	-4	10.7-10.9	35		

APPENDIX 3

THE OSI MODEL*

* Prepared by Joseph N. Pelton, Jr.

SEVEN LAYERS OF ISDN SYSTEM

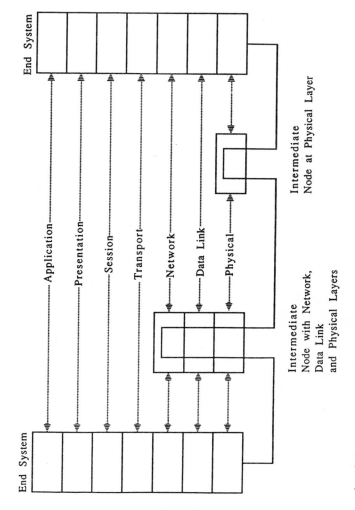

Example of OSI end systems communicating via two intermediate relay nodes

ABBREVIATIONS AND ACRONYMS

ACTS	Advanced Communications Technology Satellite
AEROSAT	Aeronautical Satellite System
AFROSAT	African Domestic Satellite System
AM	amplitude modulation
AMSC	American Mobile Satellite Corporation
AMSS	Aeronautical Mobile Satellite Service
ANIK	Canadian domestic satellite network
ANSI	American National Standards Institute
APKS	amplitude phase keyed modulation
APT	Asia Pacific Telecommunity
APTU	African Postal and Telecommunications Union
ARABSAT	Arab Satellite Communications Organization
ARIANE	European commercial launch vehicle
ASA	American Standards Association

ASETA	Association of State Telecommunications Undertakings of the Andean Sub Regional Agreement
ATA	Aeronautical Telecommunications Agency
AT&T	American Telephone and Telegraph Company
ATU	Arab Telecommunication Union
AUSSAT	Australian Satellite Network
BBC	British Broadcasting Corporation
BER	bit error rate
bps	bits per second
BRAZILSAT	Brazilian Communications Satellite
BSS	broadcast satellite service
BTI	British Telecommunications International
CARIBNET	Caribbean countries satellite network
CCIR	International Radio Consultive Committee
CCITT	International Telegraph and Telephone Consultive Committee
CCTS	Coordinating Committee on Satellite Communications
CEPT	European Conference of Postal and Telecommunications Administrations
CESA	Canadian Engineering Standards Association
CETTEM	Center of Telecommunications for the Undeveloped Countries

CFDM/FM	commanded frequency division multiplex/frequency modulation
CIRM	International Marine Radio Association
CISPR	International Special Committee on Radio Interference
COMSAT	Communications Satellite Corporation
COPUOS	Committee on the Peaceful Uses of Outer Space
CPM	Conference Prepatory Meeting
CSG	Guiana Space Center
DAMA	demand assignment multiple access
DBS	direct broadcast satellite
dBW	decibel watts
DFS	*Deutscher Fernmelde Satellite*
DGT	*Direction Générale des Télécommunications*
DOD	Department of Defense
DOMSATs	domestic satellite system
DPCM	differential pulse code modulation
DRSS	Data Relay Satellite System
DSI	digital speech interpolation
EARC	Extraordinary Administrative Radio Conference
EBU	European Broadcasting Union

ECS	European Regional Communications Satellite System
EEC	European Economic Community
EIRP	effective isotropic radiated power
EMBARTEL	Brazilian governmental telecom agency
ESA	European Space Agency
ETSI	European Telecommunication Standards Institute
EURONET	European Public Network (data)
EUTELSAT	European Telecommunications Satellite Organization
FAA	Federal Aviation Administration
FCC	Federal Communications Commission
FDM	frequency division multiplex
FDMA	frequency division multiple access
FSS	fixed satellite service
Gbps	gigabits per second
GHz	gigahertz
GSO	geostationary satellite orbit
HDTV	high definition television
HELVESAT	Swiss broadcast satellite system
IADP	INTELSAT Assistance and Development Program
IARU	International Amateur Radio Union

IBS	INTELSAT Business Services
ICAO	International Civil Aviation Organization
IEEE	Institute of Electrical and Electronic Engineers
IFRB	International Frequency Registration Board
IMCO	Intergovernmental Maritime Consultive Organization (now IMO)
IMO	International Maritime Organization (formerly IMCO)
IMTS	International Message Telephone Service
INMARSAT	International Maritime Satellite Organization
INSAT	Indian national satellite system
INTELSAT	International Telecommunications Satellite Organization
INTERSPUTNIK	International Space Telecommunication Organization
IRAC	Interdepartment Radio Advisory Committee
ISDN	integrated services digital network
ISI	International Satellites, Inc.
ISO	International Organization for Standardization
ITALSAT	Italian domestic satellite system
ITU	International Telecommunication Union
kbps	kilobits per second
kHz	kilohertz

MARECS	European Space Agency maritime satellites
MARISAT	Maritime Satellite System
Mbps	megabits per second
MCI	MCI Communications, Inc.
MHz	megahertz
MSS	mobile satellite service
MSAT X	Mobile Satellite Experiment
NASA	National Aeronautics and Space Administration
NHK	*Nippon Hoso Kyokai* (Japan)
NTIA	National Telecommunications and Information Administration
NTT	Nippon Telegraph and Telephone Corporation
OBP	on-board baseband processor
OECD	Organization for Economic Cooperation and Development
OIRT	International Radio and Television Organization
OSBS/TDMA	on board stored baseband switched time division multiple access
OSI	open systems interconnect
OTC	Overseas Telecommunications Commission
PALAPA	Indonesian domestic satellite system
PANAMSAT	Pan American Satellites

PATU	Pan African Telecommunications Union
PBX	private branch exchange
PDS	planned domestic service
PFB	Provisional Frequency Board
PFD	power flux density
PPM	pulse position modulation
PSTN	public switched telecommunication network
PTM	pulse time modulation
PTT	post, telegraph, and telephone administration
RARC	Regional Administrative Radio Conference
RCA	Radio Corporation of America
RDSS	radiodetermination satellite service
SATCOM	RCA's domestic satellite network
SBS	Satellite Business Systems
SHARE	Satellites for Health and Rural Education
SITA	*Société Internationale Télécommunications Aéronautique*
SPACENET	GTE domestic satellite network
SRG	Swiss satellite network
SS/TDMA	satellite switched time division multiple access
STW	Chinese satellite system
SYMPHONIE	French German experimental satellite system

TASI	time assigned speech interpolation
TAT	trans Atlantic telephone
TDF	French domestic broadcast satellite network
TDM	time division multiplex
TDMA	time division multiple access
TELECOM	French satellite network
TELESAT	Canadian Satellite Organization
UNESCO	United Nations Educational, Scientific, and Cultural Organization
UNISAT	British domestic satellite system
USIA	United States Information Agency
VISTA	INTELSAT thin route service
VSAT	very small aperture terminal
WARC	World Administrative Radio Conference
WESTAR	Western Union's domestic satellite
WMO	World Meteorological Organization

REFERENCES

Government Publications

INMARSAT, *Annual Review, 1987/1988*, London, 1988.

INTELSAT, *Agreement Relating to the International Telecommunications Satellite Organization,"INTELSAT,"* Washington, D.C., August 20, 1971.

_____, Board of Governors, *Annual Report of the Board of Governors to the Eighteenth Meeting of Signatories on INTELSAT Services*, Washington, D.C., 16 April 1988.

_____, *INTELSAT Annual Report 1979*, Washington, D.C., 1979.

_____, *INTELSAT Report 1986-87*, Washington, D.C., 1987.

_____, *INTELSAT Report 1987-88*, Washington, D.C., 1988. ITU, CCIR, *Report on the CCIR Conference Preparatory Meeting (CPM), Joint Meeting, Study Groups 1, 2, 4, 5, 7, 8, 9, 10, and 11, Geneva, 25 June to 20 July 1984*, Geneva, 1984.

_____, CCIR, *Technical Bases for the World Administrative Radio Conference on the Use of the Geostationary Satellite Orbit and the Planning of the Space Services Utilizing it (WARC-ORB (1)*, Geneva, 1984, Part II.

_____, *Final Acts Adopted by the Second Session of the World Administrative Radio Conference on the Use of the Geostationary-Satellite Orbit and the Planning of Space Resources Utilizing It (ORB-88), Geneva, 1988*, Geneva, 1988.

_____, *Final Acts of the Extraordinary Radio Conference for Space Communications, Geneva, 1971*, Geneva, 1971.

_____, *Final Acts of the World Administrative Radio Conference for the Planning of the Broadcasting-Satellite Service in Frequency Bands 11.7-12.2 GHz (in Regions 2 and 3) and 11.7-12.5 GHz (in Region 1), Geneva, 1977*, Geneva, 1977.

_____, *International Telecommunication Convention, Malaga-Torremolinos, 1973*, Geneva, 1974.

_____, *International Telecommunication Convention, Nairobi, 1982*, Geneva, 1983.

_____, *Radio Regulations, Edition of 1982* (Revised in 1985, 1986, and 1988), Geneva, 1985.

_____, *Radio Regulations, Geneva, 1959*, Geneva, 1960.

_____, *Report on the Activities of the International Telecommunication Union in 1987*, Geneva, 1988.

_____, World Administrative Radio Conference on the Use of the Geostationary-Satellite Orbit and the Planning of Space Services Utilizing It, First Session, Geneva, 1985, *Report to the Second Session of the Conference, Geneva, 1985*, Geneva, 1986.

OECD, *Trends of Change in Telecommunications Policy*, Paris, 1987.

University of Colorado, Center for Space and Geosciences Policy, *Report of the ACTS/Science Workshop: Potential Uses of the Advanced Communications Satellite to Serve Certain Communication, Information, an Data Needs to the Science Community*, Boulder, CO, 1988.

U.S., *Communications Satellite Act of 1962*, Public Law 87 622 Aug. 31, 1969.

U.S., Congress, Senate, Committee on Commerce, Science, and Transportation, *Long Range Goals in International Telecommunications and Information: An Outline for United States Policy*, 98th Congress, 1st Session, 1983, S. Prt. 98-22 U.S. Government Printing Office, Washington, D.C., 1983.

U.S., Department of Commerce, National Telecommunications and Information Administration, *Manual of Regulations and Procedures for Federal Radio Frequency Management, May 1986 Edition*, Washington, D.C., May 30, 1986 and Revisions for May 198, June 8, 1987.

_____, National Telecommunications and Information Administration, *NTIA Telecom 2000*, Washington, D.C., October 1988.

U.S., Department of State, *Report of the United States Delegation to the ITU Region 2 Administrative Radio Conference on the Broadcasting Satellite Service, Geneva, Switzerland, June 13 - July 17, 1983*, Washington, D.C., October 31, 1983 (mimeo.).

_____, *Report of the United States Delegation to the ITU World Administrative Radio Conference on the Use of the Geostationary Satellite Orbit and on the Planning of the Space Services Utilizing It, Geneva, Switzerland, August 29 - October 6, 1988*, Washington, D.C., March 20, 1989 (mimeo.).

_____, *Report of the United States Delegation to the World Administrative Radio Conference for Space Telecommunications, Geneva, Switzerland, 7 June to 17 July, 1971*, TD, Serial No. 26, Washington, D.C., 1971.

_____, Treaties and Other International Acts Series 7523, *International Telecommunications Satellite Organization (INTELSAT).* *Agreement Between the United States of America and Other Governments, and Operating Agreement, with Annex Concluded by Certain Governments, and Entities Designated by Governments*, Washington, D.C., August 20, 1971.

_____, Bureau of International Communications and Information Policy, *Report of the United States Delegation to the First Session of the ITU World Administrative Radio Conference on the Planning of the Geostationary Satellite Orbit and the Space Services Utilizing It, Geneva, Switzerland, August 8 - September 16, 1985*, Washington, D.C., February 21, 1986 (mimeo).

_____, Bureau of International Communications and Information Policy, *Report of the United States Delegation to the ITU World Administrative Radio Conference on the Use of the Geostationary Satellite Orbit and on the Planning of Space Services Utilizing It, Geneva, Switzerland, August 29-October 6, 1988*, Washington, D.C., March 20, 1989 (mimeo.).

_____, Office of International Communications Policy, *Report of the Chairman of the United States Delegation to the World Administrative Radio Conference of the International Telecommunication Union, Geneva, Switzerland, September 24 - December 6, 1979*, TD Serial No. 116, Washington, D.C., 1980 (mimeo.).

_____, Office of International Telecommunications Policy, *Report of the Chairman of the United States Delegation to the World Administrative Radio Conference of the International Telecommunication Union, Geneva, Switzerland, September 24--December 6, 1979*, TD Serial No. 116, Washington, D.C., 1980 (mimeo.).

_____, Telecommunications Division, *Report of the Chairman of the United States Delegation to the Extraordinary Administrative Radio Conference To Allocate Frequency Bands for Space Radiocommunication Purposes, Geneva, October 7, 1963 through November 8, 1963*, Washington, D.C., December 15, 1963 (mimeo.).

U.S., Department of State and Department of Commerce, *A White Paper on New International Satellite Systems*, Washington, D.C., February 1985 (mimeo.) 51 pp.

U.S., Federal Communications Commission, *Common Carrier Services Information: Application Filing Requirements for Domestic Satellite Earth Station Authorizations (Revised)*, Public Notice, Released: June 18, 1987, DA 87-732, Washington, D.C., 1987.

_____, *Report and Order in the Matter of Establishment of Satellite Systems Providing International Communications*, CC Docket No. 84-1299, FCC 85-399, July 25, 1985.

194

_____, *Rules and Regulations*, September, 1987.

_____, *Second Report and Order, June 16, 1972*, Docket No. 16495 (35 FCC 2nd 844).

U.S., General Accounting Office, *Federal Communications Commission Spectrum Management*, GAO/RCED-89-62, Washington, D.C., January 12, 1989.

_____, *U.S. Objectives Generally Achieved at Broadcasting Satellite International Conference -- Improvements Can Help in Future Conferences*, GAO/RECD-84-157, Washington, D.C., August 2, 1984.

Books

Apter, Joel, and Joseph N. Pelton (eds.), *The International Global Satellite System* (New York, NY: American Institute of Aeronautics and Astronautics), 1984.

Arnold, John B., *Future Satellite Systems Study* (Washington, D.C.: COMSAT Corporation), December 10, 1988.

Baker, H. Charles, John C. Bellamy, John L. Fike, and George E. Friend, *Understanding Data Communications* (Indianapolis, IN: Howard W. Sams & Co.), 1984.

Black, Uyless, *Sata Communications and Distributed Networks*, (Englewood. Cliffs, NJ: Prentice Hall, Inc.), 1987.

Bruce, Robert R., Jeffrey P. Cunard, Mark D. Director, *From Telecommunications to Electronic Services: A Global Spectrum of Definitions, Boundary Lines, and Structures* (London: International Institute of Communications), 1986.

Chetty, P.R.K., *Satellite Technology and its Applications* (Blue Ridge Summit, PA: Tab Books, Inc.), 1988.

Codding, George A., Jr., *The International Telecommunication Union: An Experiment in International Cooperation* (Leiden: E.J. Brill), 1952 (Reprinted in 1972 by Arno Press, New York).

Codding, George A., Jr. and Anthony M. Rutkowski, *The International Telecommunication Union in a Changing World* (Dedham, MA: Artech House, Inc.), 1982.

Demac, Donna A. (ed.), *Tracing New Orbits: Cooperation and Competition in Global Satellite Development* (New York, NY: Columbia University Press), 1986.

Eward, Ronald S., *The Competition for Markets in International Telecommunications* (Dedham, MA: Artech House, Inc.), 1984.

Galloway, Jonathan F., *The Politics and Technology of Satellite Communications* (Lexington, MA: D.C. Heath and Company), 1972.

Howell, W.J., *World Broadcasting in the Age of the Satellite: Comparative Systems, Policies, and Issues* (Norwood, N.J.: Ablex Publishing Corp.), 1986.

Hudson, Heather E. (ed), *New Directions in Satellite Communications: Challenges for North and South* (Dedham, MA: Artech House, Inc.), 1985.

Jansky, Donald M., *World Atlas of Satellites* (Dedham, MA: Artech House Inc.), 1983.

Kinsley, Michael E., *Outer Space and Inner Sanctums: Government, Business, and Satellite Communication* (New York, NY: John Wiley & Sons), 1976.

Leeson, Kenneth W., *International Communications: Blueprint for Policy* (New York, NY: North-Holland), 1984.

Long, Mark, *World Satellite Almanac* (Indianapolis: Howard W. Sams and Co.), 1987.

Magnant, Robert S., *Domestic Satellites: An FCC Giant Step* (Boulder, CO: Westview Press, Inc.), 1977.

Michigan Yearbook of International Legal Studies, 1984, *Regulation of Transnational Communications* (New York, NY: Clark Boardman Company, Ltd.), 1984.

Mowlana, Hamid, *Global Information and Word Communication* (New York: Longman), 1986.

Naderi, F. Michael, and S. Joseph Campanella, *NASA's Advanced Communications Technology Satellite (ACTS): An Overview of the Satellite, the Network, and the Underlying Technologies* (Washington, D.C.: NASA), March 1, 1988.

Pelton, Joseph N., and John Howkins, *Satellites International* (New York, NY: Stockton Press), 1988.

Saunders, Robert J., Jeremy J. Warford, and Bjorn Wellenius, *Telecommunications and Economic Development* (Baltimore: Johns Hopkins Press), 1983.

Savage, James G., *The Politics of International Telecommunication Regulation* (Boulder, CO: Westview Press), 1989.

Snow, Marcellus S., and Meheroo Jussawalla (eds.), *Telecommunication Economics and International Regulatory Policy: An Annotated Bibliography* (Westport, CT: Greenwood Press), 1986.

Tunstall, Jeremy, *Communications Deregulation: Unleashing America's Communications Industry* (New York, NY: Basil Blackwell), 1986.

White, Rita Lauria, and Harold M. White, Jr., *The Law and Regulation of International Space Communication* (Norwood, MA: Artech House, Inc.), 1988.

Articles and Chapters in Books

Athanassiadis, D., "The Canadian MSAT Program," *MSAT-X Quarterly*, February 1986, pp. 3-4.

Bennet, Tamara, "Private Networks 1988," *Satellite Communications*, July 1988, pp. 32-39.

Bennet, Tamara, "Prodat Enters Europe's Mobile Fray," *Satellite Communications*, March 1988, pp. 23-26.

Burch, Dean, "Intelsat: The Tomorrow Organization," in Joseph N. Pelton and John Howkins, *Satellites International*, New York, NY: Stockton Press, 1987, pp. 25-27.

Caruso, Andrea, "The European Telecommunications Satellite Organization," in Joseph N. Pelton and John Howkins, *Satellites International*, New York, NY: Stockton Press, 1987, p. 31.

Chase, Scott, "Teleports: What They Are Depends on Who's Talking," *Via Satellite*, August 1987, pp. 22-25.

Codding, George A., Jr., "The Glodom Alternative: Domestic Satellite Systems for Developing Countries," *Space Policy*, August 1989, pp. 227-236.

Contreix, Simone, "EUTELSAT: Europe's Satellite Telecommunications," in *Regulation of Transnational Communication*, 1984 Michigan Yearbook of International Studies, New York, NY: Clark Boardman Company, Ltd., 1984, pp. 85-101.

Cowhey, Peter F., and Jonathan D. Aronson, "The Great Satellite Shootout," *Regulation*, May/June 1985, pp. 27-35.

Dornheim, Michael A., "Potential Satellite Services Laboring Under Conflicting Frequency Systems." *Aviation Week and Space Technology*, January 4, 1988, pp. 261-282.

DuCharme, E.D., R.R. Bowen, and M.J.R. Irwin, "The Genesis of the 1985/87 ITU World Administrative Radio Conference on the Use of the Geostationary Satellite Orbit and the Planning of Space Services Utilizing It," *Annals of Air and Space Law*, Vol. VII, 1982, pp. 261-282.

Edelson, Burton I., "The Experimental Years," in Joel Apter and Joseph N. Pelton (eds.), *The International Global Satellite System*, New York, NY: American Institute of Aeronautics and Astronautics, 1984, pp. 39-53.

Finch, Michael J., "Limited Space: Allocating the Geostationary Orbit," *Northwestern Journal of International Law and Business*, Vol. 7, Fall/Winter 1986, pp. 788-802.

Freeman, H.L., "Telecommunication: A Necessary Partner of Business, Trade and Industry," *Telecommunications Journal*, Vol. 4, No. VII (July 1987), pp. 443-448.

Gershon, R.A., "Satellites and Fiber Optics: Redefining the Business of Long-Hall Transmission," *Business Communications Review*, Vol. 18, No. 2 (March-April 1987), pp. 22-25.

Gorove, Stephen, "Major Issues Arising from the Use of the Geostationary Orbit," in Leslie J. Anderson (ed.), *Regulation of Transnational Communications*, 1984 Michigan Yearbook of International Studies. New York, NY: Clark Boardman, 1984, pp. 3-12.

Grimes, Amy J., "Satellites, Fiber Will Coexist," *Satellite Communications*, May 1988, pp. 37-38.

Henneman, Gerhard, "Teleports: The Global Outlook," *Satellite Communications*, May 1987, pp. 29-33.

Jefferson, Patricia, and Johan Benson, "Edward C. Welsh: A Personal History of the Comsat Act," *Satellite Communications*, January 1979, pp. 22-25.

Johnson, Leland L., "Excess Capacity in International Telecommunications," *Telecommunications Policy*, September 1987, pp. 281-295.

Josentuliyan, "The Establishment of an International Maritime Satellite System," *Annals of Air and Space Law*, Vol. 2, 1977, pp. 323-349.

Jussawalla, Meheroo, "The Economic Implications of Satellite Technology and the Industrialization of Space," *Telecommunications Policy*, Vol. 8, No. 3 (September 1984), pp. 237-248.

Kalish, David, and Fred Feldman, "Fiber Optic Trends in the United States," *Conference Proceedings, Fibersat 86*, Vancouver, Canada: Westcom Communications Studies and Research Ltd., 1986, pp. 331-332.

Kerver, Tom, "North Atlantic Fury," *Satellite Communications*, August 1985, pp. 19-24.

Koehler, John E., "Satellite Communications -- International Considerations," *IEEE Communications Magazine*, Vol. 25, No. 1 (January 1987), pp. 33-35.

Leive, David M., "Some Conflicting Trends in Satellite Communications," in Leslie J. Anderson (ed.), *Regulation of Transnational Communications*, 1984 Michigan Yearbook of International Law, New York, NY: Clark Boardman, 1984, pp. 73-82.

Levin, Harvey J., "Emergent Markets for Orbit Spectrum assignments: An Idea Whose Time has Come," *Telecommunications Policy*, March 1988, pp. 57-76.

Levin, Harvey J., "Latecomer Cost Handicap in Satellite Communications," *Telecommunications Policy*, June 1985, pp. 121-135.

Levy, Steven A., "Institutional Perspectives on the Allocation of Space Orbital Resources: The ITU, Common User Satellite Systems and

198

Beyond," *Case Western Journal of International Law,* Vol. 16 (2), 1984, pp. 171-202.

Logue, Timothy J., "U.S. Decisions on Pacific Telecommunications Facilities: Letting a Million Circuits Bloom," *Jurimetrics Journal,* Fall, 1986, pp. 65-76.

Lowndes, Jay C., "U.S. Opposes Rigid Planning of Geostationary Orbit, Spectrum," *Aviation Week and Space Technology,* November 4, 1985, pp. 69-73.

Lundberg, Olof, "Inmarsat on the Move," in Joseph N. Pelton and John Howkins (eds.), *Satellites International,* New York, NY: Stockton Press, 1987, pp. 29-30.

Maherzi, Lofti, "A Highway Into Space," *Intermedia,* Vol. 24, No. 6 (November 1986), pp. 19-21.

Mason, Charles, "Circuit Loading Guidelines Out," *Telephony,* March 28, 1988, p. 3.

Matsumoto, Shinji, "Present and Future Domestic Satellite Communication Systems in Japan," *Conference Proceedings, Fibersat 86,* Vancouver, Canada: Westcom Communications Studies and Research Ltd., 1986, pp. 43-49.

McDougal, P.J., and Joseph N. Pelton, "ISDN: The Case for Satellites," *Telecommunication Journal,* Vol. 54, No. V (May 1987), pp. 318-322.

McKenna, Julianne, "Bypassing Intelsat: Fair Competition or Violation of the Intelsat Agreement," *Fordham International Law Review,* Vol. 8, No. 3 (1984-85), pp. 479-512.

McKnight, Lee, "Implications of Deregulating Satellite Communications," *Telecommunications Policy,* December 1985, pp. 277-280.

McLernon, Lawrence A., "Fiber Optics as a Replacement Technology," in *Conference Proceedings, Fibersat 86,* Vancouver, Canada: Westcom Communications Studies and Research Ltd., 1986, pp. 345-349.

Mendel, Sigrid Arlene, "Authorization of Private International Satellite Systems in Competition with COMSAT: An Analysis of the Underlying Legal Justification and Policy Factors," *Law and Policy in International Business,* Vol. 18, 1986, pp. 279-311.

Murphy, Barry, "Satellites in Canada: Past, Present, and Future," *Conference Proceedings, Fibersat, 86,* Vancouver, Canada: Westcom Communications Studies and Research Ltd., 1986, pp. 83-85.

Naraine, Mahindra, "WARC-ORB-85: Guaranteeing Access to the Geostationary Orbit," *Telecommunications Policy,* June 1985, pp. 97-108.

199

Narjes, Karl-Heinz, "Towards a European Telecommunication Community: Implementing the Green Paper," *Telecommunications Policy*, June 1988, pp. 106-108.

Parker, Edwin D., "Micro Earth Stations as Personal Computer Accessories," *Proceedings of the IEEE*, November 1984, pp. 1526-1531.

Pelton, Joseph N., "A User-Friendly Introduction to Satellites," in Joseph N. Pelton and John Howkins, *Satellites International*, New York, NY: Stockton Press, 1987, pp. 1-3.

_____, "The Future of Telecommunications: A Delphic Survey," *Journal of Communication*, Vol 32, No. 1 (Winter 1981), pp. 177-189.

Pelton, Joseph N., and P.J. Mcdougal, "ISDN: The Case for Satellites," *Telecommunications Journal*, Vol. 54, No. VII (July 1987), pp. 317-312.

Pelton, Joseph N. and William W. Wu, "The Challenge of 21st Century Satellite Communications: Intelsat Enters the Second Millennium," *IEEE Journal on Selected Areas in Communications*, Vol. SAC-5, No. 4 (May 1987), pp. 571-591.

Pirard, Theo, "EUTELSAT: Satellites Linking Europe," *Satellite Communications*, July 1983, pp. 34-42.

_____, "INTERSPUTNIK: The Eastern Brother of INTELSAT," *Satellite Communications*, August 1982, pp. 38-44.

_____, "New Markets for the New INMARSAT," *Satellite Communications*, March 1988, pp. 20-22.

Potamitis, Stathis, "Competition in International Satellite Telecommunications Services," *University of Toronto Faculty of Law Review*, Vol. 44, No. 1 (Spring 1986), pp. 33-56.

Pritchard, Wilbur, "The Basics of Satellite Technology," in Joseph N. Pelton and John Howkins, *Satellites International*, New York, NY: Stockton Press, 1987, pp. 19-24.

Rivera, Henry M., "Separate Systems and International Communications," *IEEE Communications Magazine*, Vol. 25, No. 1 (January 1987), pp. 39-43.

Rothblatt, Martin A., "ITU Regulation of Satellite Communication," *Stanford Journal of International Law*, Vol. 18, September 1982, pp. 1-24.

_____,"Radiodetermination Satellite Service," *Telecommunications*, June, 1987, pp. 39-42.

_____, "The Impact of International Satellite Communications Law upon Access to the Geostationary Orbit and the Electromagnetic Spectrum," *Texas International Law Journal*, Vol. 16, 1981, pp. 207-244.

Runge, Peter K., and Patricia Trishitts, "Future Underseas Light Wave Communications Systems," *Signal*, June 1983, pp. 30-35.

200

Rutkowski, Anthony M., "Beyond Fiber Optics vs. Satellites," *Telecommunications*, September 1985, pp. 11-15.

_____, "Six Ad-Hoc Two: The Third World Speaks its Mind," *Satellite Communications*, March 1980, pp. 22-27.

Schwartz, Irwin B., "Pirates or Pioneers in Orbit? Private International Communications Satellite Systems and Article XIV(d) of the Intelsat Agreement," *Boston College International and Comparative Law Review*, Vol. IX, No. 1 (Winter 1986), pp. 199-242.

Scott, John Carver, "Back in the USSR," *Satellite Communications*, April 1988, pp. 43-46.

Smith, Milton L., "Space WARC 1985: The Quest for equitable Access," *Boston University International Law Journal*, Vol. 3, No. 2 (Summer 1985), pp. 229-255.

Snow, Marcellus S., "Arguments For and Against Competition in International Satellite Facilities and Services: A U.S. Perspective," *Journal of Communications*, Vol. 35, No.3 (September 1985), pp. 51-79.

_____, "Natural Monopoly in INTELSAT: Cost Estimation and Policy Implications for the Separate Systems Issue," *Telematics and Informatics*, Vol. 4, No. 2 (1987), pp. 133-150.

_____, "The Economics of Satellite Communication," in Joseph N. Pelton and John Howkins (eds.), *Satellites International*, New York, NY: Stockton Press, 1988, pp. 49-55.

Solomon, J., "PanAmSat's Proposed Entry Into the UK," *Telecommunications Policy*, September 1988, pp. 206-207.

Solomon, Jonathan, "The Future Role of International Satellite Institutions," *Telecommunications Policy*, September, 1984, pp. 213-221.

Soroos, Marvin S., "The Commons in the Sky: The Radio Spectrum and Geosynchronous Orbit as Issues in Global Policy," *International Organization*, Vol. 36, No. 3 (Summer, 1982), pp. 665-677.

Staple, Gregory C., "The New World Satellite Order: A Report from Geneva," *American Journal of International Law*, Vol. 80, No. 3 (July 1986), pp. 699-720.

Steele, L.C., "MSS Update," *MSAT-X Quarterly*, No. 21, October 1989, pp. 1-3.

Stern, Martin L., "Communication Satellites and the Geostationary Orbit: Reconciling Equitable Access with Efficient Use," *Law and Policy in International Business*, Vol. 14(3), 1982, pp. 859-883.

Stoddard, Rob, "Decade of Deregulation," *Satellite Communications*, July 1987, pp. 21-22.

_____, "Regulatory Outlook '89," *Satellite Communications*, January 1989, pp. 19-22.

201

____, "Rethinking International Satellite Regs," *Satellite Communications*, September 1987, pp. 37-39.

The Georgetown Space Law Group, "The Geostationary Orbit: Legal, Technical and Political Issues Surrounding Its Use in World Telecommunications," *Case Western Reserve Journal of International Law*, Vol. 16(2), 1984, pp 223-264.

Tucker, Elizabeth, "Comsat Signs Contract to Ensure AT&T Traffic," *Washington Post*, October 10, 1987.

Valovic, Tom, "Fiber-Optic Deployment Along the Interchange Carriers," *Telecommunications*, May 1987, pp. 40-53.

Waite, Barbara L., and Ford Rowan, "International Communications Law, Part II: Satellite Regulation and the Space WARC," *The International Lawyer*, Vol. 20, No. 1 (Winter 1986), pp. 341-365.

Warr, Michael, "Kessler Projects FO Expansion," *Telephony*, June 27, 1988, pp. 13-14.

Withers, D., "Equitable Access to Satellite Communication," *Electronics & Wireless World*, December 1985, pp. 65-66 & 75.

Wheelon, Albert D., "The Future of Communication Satellites," *Intermedia*, Vol. 12, No. 2 (March 1984), pp. 40-49.

Other

Smith, Milton S., III, *Space WARC 1985 -- Legal Issues and Implications*, an LL.M thesis submitted to the Faculty of Graduate Studies and Research at the Institute of Air and Space Law of McGill University, 1984.

INDEX